MW01688076

THE TRUTH ABOUT CB ANTENNAS

William I. Orr, W6SAI

Stuart D. Cowan, W2LX

RADIO AMATEUR CALLBOOK

P.O. BOX 2013 LAKEWOOD, NEW JERSEY 08701

Acknowledgments: We wish to thank the following companies and organizations for their kind assistance in providing photographs, drawings, and technical information used in the preparation of this book: American Radio Relay League, Inc.; Antenna Specialists Company; Avanti Research and Development, Inc.; Cush-craft Corporation; Hy-Gain Electronics Corporation; E. F. Johnson Company; Kirk Electronics, a division of Electrotec Corp.; Mosley Electronics, Inc.; New-Tronics Corporation; Raytheon Company; and The Shakespeare Company. These organizations bear no responsibility for the opinions expressed by the authors.

Caution: Working on antennas or towers can be dangerous. All warnings on the equipment and all operating and use instructions should be adhered to. Make sure that the antenna is disconnected from the station equipment before you begin to work on it. Make sure that your antenna is not close to power lines, and that it cannot drop on a power line if wires or supports fail. Do not attempt to climb a tower without a safety belt. It is best to work on your antenna with someone who can assist you with tools and will be able to help in the event of a problem or an emergency.

Published in 1993 by Radio Amateur Callbook
(an imprint of Watson-Guptill Publications,
a division of BPI Communications, Inc.),
P.O. Box 2013, Lakewood, New Jersey 08701

Library of Congress Catalog Card Number: 70-164932
ISBN 0-8230-8708-5

Manufactured in the United States of America

1 2 3 4 5 6 7 8 9/01 00 99 98 97 96 95 94 93

TABLE OF CONTENTS

CHAPTER 1. SUGAR-COATED ANTENNA FUNDAMENTALS . . 5
(What You Don't Know *Will* Hurt You!)

CHAPTER 2. RADIO WAVES AND THE NATURE OF THINGS . 12
(Old Mother Nature's Mysterious Ways)

CHAPTER 3. YOUR ANTENNA AND SIGNAL INTERCEPTION. 21
(The Antenna and Radio Propagation)

CHAPTER 4. THE CURIOUS ANTENNA 32
(Will It Ever Replace Smoke Signals?)

CHAPTER 5. EVERYTHING YOU'VE ALWAYS WANTED TO KNOW ABOUT CB ANTENNAS 42
(But Were Afraid to Ask!)

CHAPTER 6. MORE ANTENNA GAIN FOR YOU! 53
(The Yagi and Quad Beams)

CHAPTER 7. THE TRUTH ABOUT ANTENNA GAIN 66
(P.T. Barnum is Alive and Well
-and Writing Advertisements for CB antennas)

CHAPTER 8. THE TRUTH TABLE 78
(Things Your Mother Never Taught You
About CB Antennas)

CHAPTER 9. THE COAXIAL CABLE 87
(Your Radio Lifeline)

CHAPTER 10. THE SWR METER 102
(Friend or Foe?)

CHAPTER 11. YOUR MOBILE ANTENNA 111
(Poor? Better? Best?)

CHAPTER 12. CB ANTENNAS FOR MARINE SERVICE 135
(Sea-B Radio on the Bounding Main)

CHAPTER 13. CB ANTENNAS YOU CAN BUILD 149
(Construction Projects for the CBer)

CHAPTER 14. BUILD A QUAD OR YAGI BEAM FOR CB 164
(Build Your Beam and Save Money!)

CHAPTER 15. BUILD A MONSTER QUAD BEAM FOR CB 187
(The King of the CB Antennas)

CHAPTER 16. CB ROUND-UP ON TEN IMPORTANT SUBJECTS 196
(Baluns, TVI, Lightning Protection and More)

GLOSSARY OF ANTENNA TERMS 214
(Say What You Mean - Mean What You Say!)

INDEX 222

FOREWORD

Citizens Band (CB) service centers and maintenance shops across the nation report that only about 10 percent of today's CB stations are truly effective, and that many of these experience communications difficulties much of the time.

What causes this shocking situation? Crowded bands and illegal operations are part of the reason but much of the trouble lies with poor antennas. The simple truth is that your CB antenna is the key to effective communication, and most CB antennas do not radiate and receive signals with anything approaching peak efficiency. Your station is only as good as your antenna; a good antenna can make a good station truly effective. The results are stronger signals and more satisfying contacts.

Your legal CB transmitter power is very low--about the power used to light two flashlight bulbs. Make the most efficient use of this low power and you'll enjoy better communications.

This Handbook makes your CB station work better by giving you the information you need to install an efficient antenna and adjust it properly--information the experts know and use to advantage. Whether you buy or build your antenna, this Handbook helps you understand antennas and tells you how to make them work for you. It helps you analyze antenna advertisements and sweep aside the inflated claims and confusing "facts" sometimes touted by hard-sell manufacturers in an effort to win your dollars.

This easy-to-understand Handbook:

- Reveals *true* CB antenna power gain in a novel "Truth Table"
- Gives you construction data for popular antennas, including the Monster Quad beam--the legal "King of CB Antennas"
- Gives you the expert's way to make your mobile CB antenna radiate maximum power
- Shows you how to make your base station antenna operate properly--and save money, too
- Explains basic antenna theory in clear, understandable terms

In short, this Handbook tells you the TRUTH about CB antennas!

Chapter 1

Sugar-Coated Antenna Fundamentals

(What You Don't Know Will Hurt You!)

Let's not take up too much time on dry, uninteresting antenna theory! You're interested in the facts and what you can do to improve your CB radio performance. You're being asked to lay out a lot of money these days for fancy antennas having fancy words said about them. Its true that antenna type and installation make a *big* difference in the performance of your CB station. But you'll find, as have others, that a good working knowledge of antennas helps you solve many of your problems, aids you in making a sensible choice when you buy an antenna and--in the long run--gives you a stronger signal.

Can you really believe what you read in the antenna advertisements? How much is true? How much is baloney? What does the advertisement mean when it talks about "signal gain," or "gain over isotropic," or "effective talk-power gain"? Are these terms meaningful?

You might be fooled , and being fooled costs you time and money. This Handbook keeps you from being fooled! This book explains terms, discusses antenna types and gives you the information you need when you go to buy or build your CB antenna.

Let's start at the beginning. Antennas have had an interesting history over the years and a study of the past may prevent errors in the future. So here's a quick, concise background of the history of radio and a short introduction to radio waves and the nature of things. We'll look at early antennas, the sugar-coated theory of radio transmission, antenna gain, and so on. It's a painless introduction, and it helps you understand how today's CB antennas work. No math, no gobbledygook. Just the straight dope.

The Early Days

Over 150 years ago experimenters found that when an electric current flowed through a wire, a magnetic field was found wrapped about the wire. Shortly thereafter, it was discovered that a *changing* magnetic field produced a flow of current in a nearby wire. From these two discoveries, Michael Faraday of England proposed the novel idea of a magnetic "flux field" or "lines of force" (invisible lines of tension in space: like stretched rubber bands) to explain the phenomenon of magnetic fields and force acting at a distance. Faraday, moreover, expanded his curious idea into a general "field theory" of force which proposed that all space was filled by various force fields: magnetic, electric, gravitational, and so on.

About 1850 a canny Scotsman, James Clerk Maxwell, derived a breath-taking concept of nature and revealed a striking set of mathematical rules that encompassed all known electromagnetic knowledge and in the broader sense predicted an entirely new theory of electromagnetic radiation, described by Maxwell in terms of his "field equations." Maxwell boldly stated that "light consists of undulations of the ether" and predicted that electric and magnetic phenomena were similar to light and that electric "undulations" could exist in free space, in the same manner as light waves. The electric "undulations" were described by a set of monumental equations that showed a wave freely travelling from place to place, with an interchange of energy constantly radiating outwards from the source.

This was quite an idea for 1850!

Hertzian Waves

Thirty-two years after Maxwell's amazing electromagnetic theory and forty years after Faraday's original suggestion of an electric field, Heinrich Hertz of Germany proved its existence. He built a powerful radio oscillator using a "sparking coil," Leyden jars and a simple antenna. The transmitter worked at a frequency of about 53 MHz (megahertz), just about where TV channel 2 is today! For a receiver, Hertz used a length of wire bent into a loop and having "sparking balls" at the gap (Figure 1). By painstaking adjustment of the gap, Hertz made his simple receiver sensitive enough to spark at a distance of about 30 feet from the oscillator. This amazing fellow then proceeded to focus his electromagnetic waves with simple directional (beam) antennas and reflect the waves from metal surfaces! Other experimenters duplicated Hertz's gear and soon extended the range of the sparking-ball receiver and sparking-coil transmitter up to three hundred feet or so.

An interested observer of these early experiments was Guglielmo Marconi, of Italy. In 1895 he started his famous experiments, culminating in his historic trans-Altantic radio transmissions in 1901. "Wireless" had come of age and by World War I, Marconi radio equipment was placed in service by various countries and used to handle messages over hundreds of miles. Contrary to the work of Hertz, early

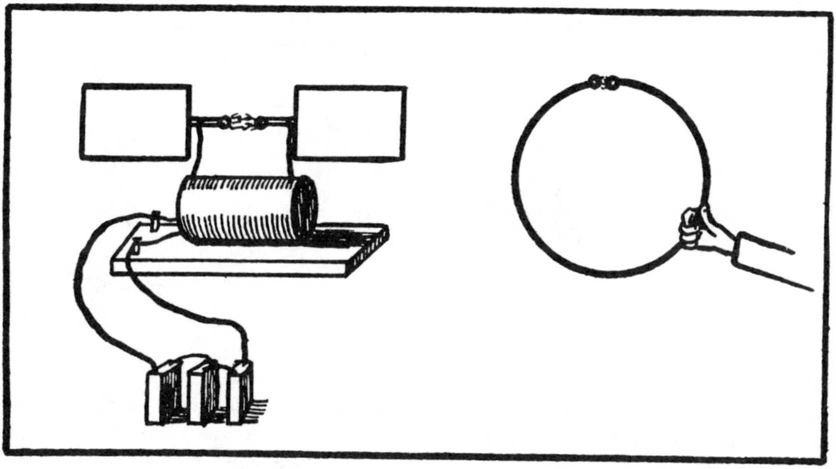

Fig. 1 **WORLD'S FIRST** radio transmitter (left) and radio receiver (right). In 1884, Heinrich Hertz of Germany generated and detected radio waves using this equipment. The waves came from a spark transmitter which used two copper plates as an antenna. Holding a resonant length of wire bent into a loop, Hertz moved about his laboratory and found that a small spark would jump across the gap in the loop within certain distances from the transmitter. In this fashion Hertz verified the formulas of Maxwell and proved radio waves existed and determined their wavelength. He also found that he could reflect and refract his waves with large metal sheets serving as a "radio mirror". (Drawing adapted from "Radio Theory and Operating", by Loomis, 1925).

"wireless" stations used very long waves, since it was obvious to these early experimenters that long waves were needed to cover long distances!

The Great Days

After World War I, the first radio amateurs experimented with "short" radio waves and frequencies as high as today's CB channels were tried by 1925. A great expansion of communication activity into the shortwave (high frequency) radio spectrum occurred when it was found that the waves could be sent around the world by bouncing them off an ionized layer of the atmosphere and back to earth. This layer (the *ionosphere*) varies in height 100 to 250 miles above the earth. With the perfection of stable frequency control for transmitters and

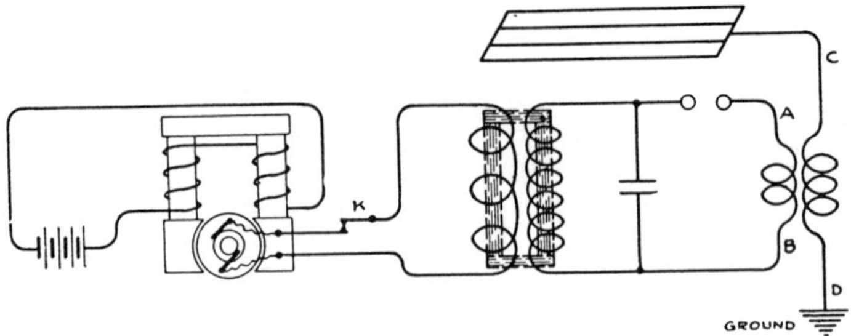

Fig. 2 SPARK RADIO TRANSMITTER of 1910 was run from a large wet battery and motor generator (left). A step-up transformer (center) provided high voltage for the spark gap and oscillation transformer (A-B). A huge cage antenna provided a communication range of 50 miles or so, using Morse code. (Drawing adapted from "Robison's Manual of Radio Telegraphy", 1918).

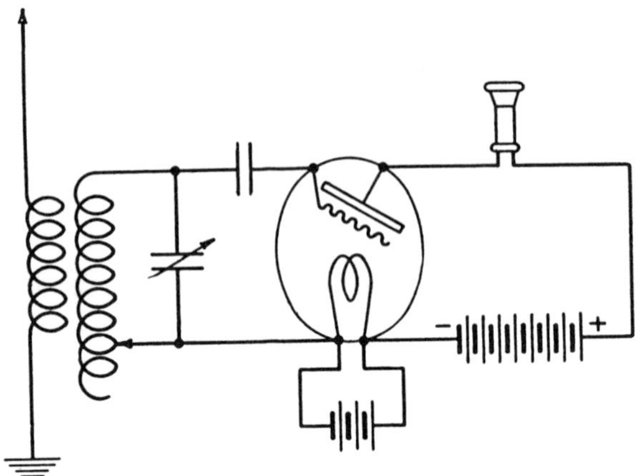

Fig. 3 FIRST RADIO SETS were simple crystal detectors. By 1920 a three element vacuum tube was used in conjunction with a telephone receiver and large antenna for reception of code up to 100 miles or so. Fragile, gassy tubes cost up to six dollars and had short life. By 1923 music broadcasting popularized radio reception. (Adapted from "Robison's Manual of Radio Telegraphy", 1918).

Fig. 4 GRAND-DADDY TRANSCEIVER of 1935 is ancestor of today's efficient CB transceiver. A single double triode tube, operating from dry cells was used. Whip antenna provided an operating range of about a mile. (Adapted from "Radio" magazine, July, 1936).

the sensitive superhetrodyne receiver about 1935, short wave radio transmission became a reality rather than a hit-or-miss proposition. Soon thereafter, the great technological explosion brought about by World War II opened the way for today's communication miracles: radar, television, signals bounced off the moon and Mars, and--finally--the transistor.

All of these modern communication techniques and devices make use of the mysterious "undulations" noted by Faraday so long ago and first put to use by Heinrich Hertz. Of the true nature of the radio wave, however, nothing is known. The radio wave, then, must not be thought of as a thing, but as a way in which things behave. For after a description is given of the behavior of the mysterious radio wave, nothing more can be said, as the ultimate knowledge of the radio wave is a secret locked in the heart of the universe.

Today's radio operator stands upon the shoulders of giants. Even today, radio is still in a formative phase, and Citizens Band Radio in

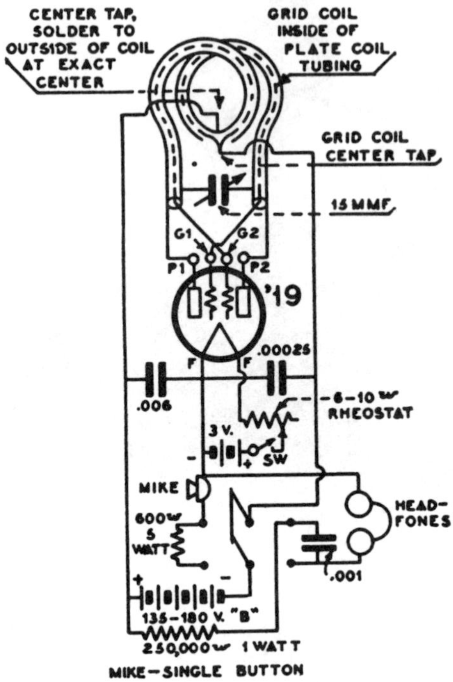

Fig. 5 ONE TUBE amateur transceiver of the "thirties" was used by radio hams in the old 6 meter band. Using a minimum of parts, the circuit was nearly fool-proof and provided voice communication for a few dollars. Circuit is outmoded and illegal to use today. (Adapted from "Radio" magazine, July, 1936).

particular is less than two decades old. Antennas, and CB antennas to boot, are still primitive, even though the theory behind them is complex. Today's antenna will probably make the CBer of 2001 smile as he shoots off to Paris on a supersonic jet with his space-approved CB transceiver and his pocket antenna, talking to CBers as he zips through space!

The Radio Antenna

Heinrich Hertz had a very clear notion of Maxwell's mathematical concept of a radio wave: an interchange of energy in space between free electric and magnetic fields, with the energy radiating outwards from the source as ripples spread out from a stone cast into a quiet pond. The practical problem was to construct a "launching device" that would project the radio wave into space to commence its journey Today, such a device is called an *antenna* . Simply speaking, any structure that radiates and intercepts radio waves is an antenna.

Fig. 6 THIS 85-FOOT ANTENNA is located at a COMSAT earth station and is used to relay high speed communication data between North America and the Far East via the Pacific satellites. This is one of six U.S. earth stations used to handle international communications. Imagine how a comparable dish antenna like this would work on the CB channels! Any takers?

Antennas come in all shapes and sizes, from the compact antenna concealed in a transistor radio to the huge towers of the modern television transmitter. Observers have noted rotating disc-shaped radar antennas on boats, backyard television antennas, and the large beam antennas of radio amateurs. Each antenna is shaped and built to do a particular job, but the task of all antennas is to launch and intercept radio waves in the manner most efficient to the task at hand. Hertz, if he were alive today, would understand and appreciate the most complex of modern communication antennas, as the design is based upon laws of nature that are inviolable.

Chapter 2

Radio Waves and The Nature of Things

(Old Mother Nature's Mysterious Ways)

The earth and everything on it is continuously bombarded by energy from outer space in the form of waves coming from countless sources. Some waves are useful (sunlight) and some may be dangerous (x-rays). Others are not fully understood (cosmic rays). Many such waves have been recruited in the service of man.

The waves from space arrive helter-skelter but the waves, or groups of waves, are orderly and have their own characteristics. Taken as a whole, when arranged by size, or *wavelength,* they make up the *electromagnetic spectrum* (Figure 1).

All of these waves, constantly in motion about us, vibrate in typical waveform at definite frequencies (the number of complete waves per second; *one cycle per second* is called *one Hertz*), with the wavelength decreasing as the frequency increases.

Wave Bands

About 1880, experimenters developed precise instruments for measuring light waves and by 1900 it was possible to measure the wavelength of radio waves, x-rays and cosmic rays. As the various waves were investigated and measured, they were grouped into *wave bands,* which are groups of waves with similar characteristics, measured in a similar fashion.

The next step was to generate or procure a supply of particular waves for study. Light waves were easy to procure. If a beam of red light was desired, for example, all that was needed was sunlight and a red filter which would block out the unwanted wavelengths and pass only red light. On the other hand, x-rays and radio waves are generated by nature in far-off space, but must be produced on earth as the space waves are too weak to be put to use.

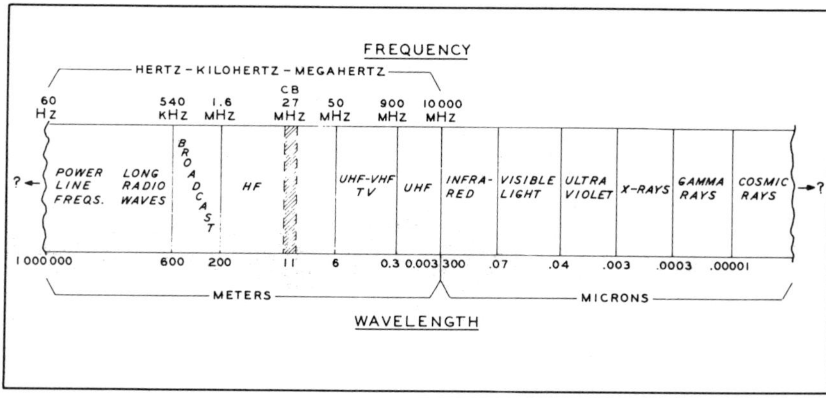

Fig. 1 THE ELECTROMAGNETIC SPECTRUM is a chart showing the relationship between electromagnetic waves arranged by size (wavelength). At the low frequency end of the spectrum are the extremely large radio waves useful for long distance daylight communication. Huge antennas are required to radiate these big waves. The regular broadcast band encompasses medium size waves and still shorter (smaller) waves are useful for long distance, ionospheric-reflected radio communication. The CB channels fall in this range at about 11 meters wavelength. Very short radio waves are used for radar and point-to-point short range communication.

Gradually, the extremely short radio waves blend into infrared waves, and at still shorter wavelengths, the electromagnetic waves are visible to the eye as light waves. Shorter than light are ultraviolet waves, X-rays, gamma rays and cosmic waves. The size of these tiny waves is expressed in terms of millionths of an inch. Even so, they are identical to the larger radio waves used by CBers.

What waves, if any, exist beyond the ends of the electromagnetic spectrum chart? Scientists suspect that super-long waves may be found in the form of gravitational waves and that micro-miniature waves shorter than cosmic waves exist in the universe. No one is sure about super-long or super-short waves and the puzzle is to find them, as man seeks to learn more about the mysterious universe and all the electromagnetic waves that surround us.

High frequency electric energy can be generated by a radio transmitter and converted into electromagnetic waves by an antenna. The field set up about the antenna is transmitted through space at the speed of light (186,000 miles per second) and may be captured by a second antenna, intercepting the wave and converting it back to electric energy, capable of being detected by a radio receiver.

In other wave bands, the generation and detection of waves is costly, complex and little-known other than by the specialists who make use of the waves. For example, gamma rays are useful for radium therapy,

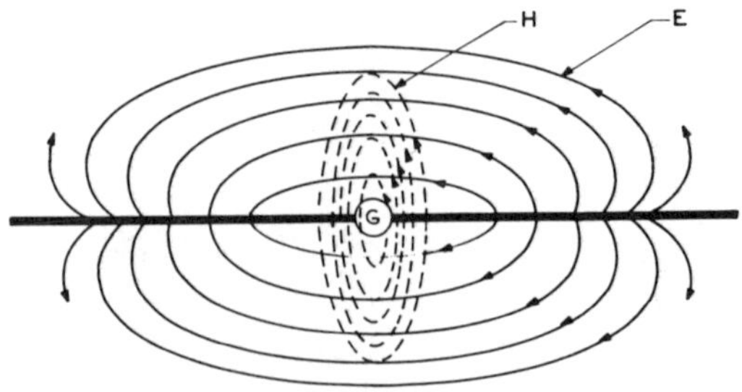

Fig. 2 MODERN CONCEPT OF A RADIO WAVE is a combination of magnetic (H) and electric (E) fields set up about an antenna as a result of electric current flowing within the antenna. Energy is transferred back and forth from one field to the other. This action is termed "oscillation". During energy transfer, which can occur millions of times per second, fields may become detatched from the wire and move off into space. A radio signal is apparently made up of the two fields which reinforce each other, with the electromagnetic energy radiating outwards from the antenna. The sum of the two fields is called an "electro-magnetic field". If the field cuts another conductor (antenna), some of the energy in the field will set electrons in motion in the conductor. The electron movement (current) may be detected by a radio receiver.

x-rays for medical and industrial applications, ultra violet waves for lasers and black light, infra red waves for drying and photography, microwaves for radar--to name a few.

Of all the wave bands, however, the radio waves have received the most use. This Handbook covers their launching and interception by antennas, and how this knowledge can best be put to use by CB operators.

The Two-Way Antenna

It was discovered early in the game that an antenna acts in the same fashion whether it is transmitting or receiving energy. Thus, the general characteristics of the antenna are the same in either instance. For simplicity, then, we can examine or discuss a particular antenna in terms of transmission or reception, with the assurance that the observations noted in one case apply to the other. The basic theory regarding similar receiving and transmitting characteristics is called *reciprocity* and, as applied to antennas, was first stated by Lord Rayleigh of England in 1877, long before radio transmission was generally known. His basic theory is still true today.

Antenna Resonance

Electrical energy is radiated into space and retrieved from space by an antenna. An antenna is a length of metal that conducts electricity and is so located that it is surrounded by space. High frequency electric energy flowing in the antenna sets up an electromagnetic field about the conductor which expands into space at the speed of light. Conversely, an electromagnetic field meeting an antenna sets up a high frequency electric current within the antenna. Thus, *any* conductor of electricity functions as an emitter or receptor of electromagnetic waves, with a varying degree of efficiency(Figure 2).

For highest efficiency, the antenna must bear some relationship to the length of the radio wave. In the early days of radio, wire fences were often used for reception and crude "cage" antennas of many wires were used for transmission without any real concept of trimming or

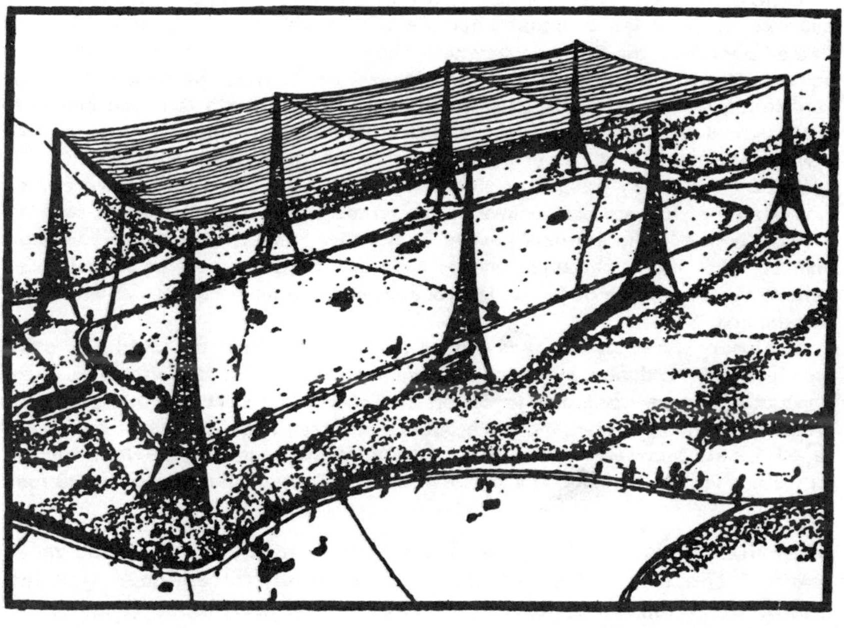

Fig. 3 WORLD'S BIGGEST ANTENNA? Huge antenna built in Lafayette, France, in 1917 for super-power long-wave station (which was never finished). The antenna covered several hundred acres and was supported on 600 foot high towers. Early radio books give vivid descriptions of big radio antennas, but operation and theory was often vague and misleading. (Adapted from "Radio Theory and Operating", by Loomis, 1925).

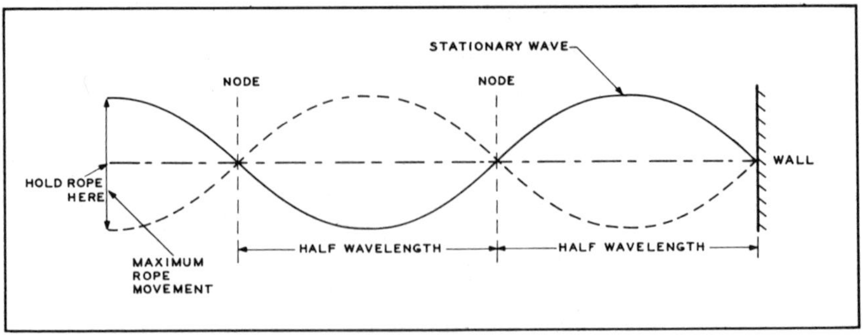

Fig. 4 RADIO WAVELENGTH. A simple case of wavelength brought about by the reflection of a wave can be observed when a rope is given a quick series of flips. A wave travels along the rope until it reaches the far end, from where it travels back along the rope to your hand, continuing back and forth in this manner until the motion dies out. By flipping the rope in the proper sequence, you can make the wave continue to run back and forth along the rope.

With a little practice, you can flip the rope in the proper sequence and a succession of waves at equal intervals will travel along the rope. When reflected back from the far end, they meet the oncoming waves whose lengths are equal to those waves coming from the far end of the rope. At some points, the conflicting waves reinforce each other and at other points the rope seems to not move at all. Points of zero movement are found along the rope one-half wavelength apart; at all other points the rope vibrates. The vibratory pattern is called a "stationary wave", or "standing wave", as the overall wave pattern moves neither forward nor backward. The points of no movement of the rope are called "nodes". The stationary wave on the rope is trapped between your hand and the other end of the rope, and by experimenting, you can get various numbers of standing wave nodes on the rope, depending upon the rate at which you flip the end.

Stationary waves of this type may be set up in an electrical circuit, or along an electrical conductor by electrical impulses applied to the circuit or to the conductor. Such a conductor is called an "antenna". Shown in this picture is an antenna having a standing wave on it, with three nodal points. A half-wavelength exists between any two nodal points. In the case of a CB radio wave of 11 meters length, the distance between adjacent nodal points is about 18 feet.

otherwise adjusting the length of the antenna to the length of the radio wave. "The bigger the antenna, the better the reception," was the motto. Such antennas, of course, were very inefficient (Figure 3).

A great body of literature exists today attesting to the many varied and important contributions made by numerous experimenters in the quest for better and more efficient antenna systems. Early in the game, it was found that an antenna of a certain length did not exhibit the same characteristics at all frequencies. Best results were obtained when the antenna was adjusted physically or electrically to be in *proportion to the length of the radio wave* . If the antenna was shorter than the wave, extra wire could be added in the form of a coil to make the antenna electrically *longer*. If the antenna was too long, a capacitor could be added in series with the antenna to make it electrically *shorter*. The antenna and its auxiliary *tuner* could thus be adjusted to the condition of *resonance*, or electrical compatibility with the radio wave in use (Figure 4).

Resonance and Antenna Length

For any antenna there is one frequency, called the *resonant frequency*, at which various characteristics of the antenna are in a state of electrical balance, and at which frequency the antenna is in a condition of maximum efficiency. The resonant frequency is a function of the *electrical length* of the antenna, which may or may not bear a relationship to the physical length in feet and inches. Any antenna may be tuned to resonance by auxiliary gadgets, but such devices may be a nuisance and of questionable efficiency. A resonant antenna requires no such devices and is a simple and effective radiator and receiver of radio energy. The length of the radio wave and the antenna is expressed in terms of *wavelength*, and that term is directly related to the *frequency* of the radio wave, as we shall see shortly.

Wavelength and Frequency

Radio waves exist because it takes a certain amount of time for electrical energy to travel from point to point. When a pebble is dropped into water, the resulting disturbance does not reach the edge of the pool immediately. Rather, a wave of water starts out from the place where the pebble hits and proceeds towards the edge at a definite speed. Electrical energy normally travels at 186,000 miles per second, or 300,000,000 meters per second. Thus, if a radio antenna is emitting a pulse of radio energy at a rate of 27,000,000 pulses per second (27,000 KHz--the CB region), one pulse will travel about 11 meters (36 feet) before another pulse is emitted. At 186,000 pulses per second, on the other hand, the distance between pulses is about 5000 feet. At 3,000,000,000 pulses per second (3,000 gigacycles), the distance between pulses is only about 4 inches. This corresponds to a radar signal. Thus, exactly as in the case of the pebble dropped into water, a mathematical relationship exists between the distance a radio pulse will travel during the time required for one pulse (one cycle) to occur, as shown in Figure 5.

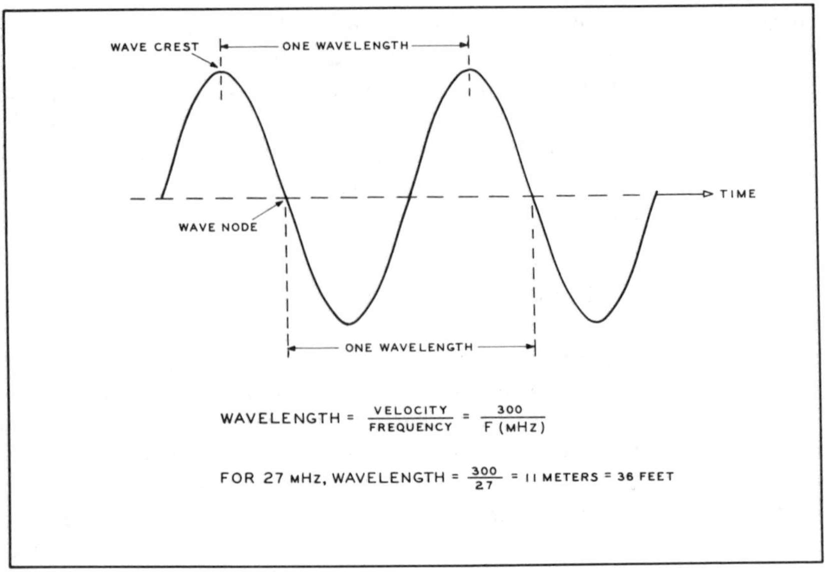

Fig. 5 WAVELENGTH AND FREQUENCY are related to each other by a simple formula, as explained in the text. The wavelength is the distance between two wave crests (or between two wave nodes). The velocity (speed) of the electromagnetic wave is 300,000,000 meters per second, the same as the speed of light. The frequency, or repetition rate, of the wave is the number of waves which pass a given point in one second. Wavelength is measured in units called "meters", one meter being equal to 3.28 feet. The frequency is measured in "cycles per second", or units called "Hertz". Thus, one cycle per second is one Hertz.

Since wavelength and frequency have an inverse relationship, it follows that long waves have a low frequency and short waves have a high frequency.

The length of any electromagnetic wave can be found by the formula:

$$\text{Wavelength in meters} = \frac{300{,}000{,}000}{\text{frequency in cycles per second}}$$

In the case of CB radio, the wavelength is expressed more conveniently as:

$$\text{Wavelength in meters} = \frac{300}{\text{frequency in megahertz (MHz)}}$$

The illustration tells the story better than words.

Measurements in the United States are usually expressed in feet and inches rather than in meters (and centimeters). This formula, converted into more familiar terms, then, is:

$$\text{One wavelength, in feet} = \frac{984}{\text{frequency in megahertz (MHz)}}$$

A half-wavelength is determined by dividing the formula by 2, thus:

$$\text{One-half wavelength, in feet} = \frac{492}{\text{frequency in megahertz (MHz)}}$$

This is the fundamental formula from which many significant lengths in antenna work are developed and is defined as the *electrical length* of a half-wave antenna element, when no factors exist that modify the speed of the radio wave. In real-life situations, many factors exist that alter the *physical length* of a half-wave antenna, or any conductor used for an antenna. If this were not true, there would be no need for this Handbook!

End effect and *thickness* of the antenna element must be taken into account when determining the actual physical length of any antenna element. The end effect is caused by the presence of insulators or other material that may be used to support the antenna at the ends and also by the abrupt transition from conductor to the surrounding atmosphere at this critical point.

Since practical antennas have thickness as well as length, the actual length departs to a small degree from the electrical length. Generally speaking, the larger in diameter the antenna element, the greater the departure from the electrical wavelength. Thus, for a given frequency an antenna having a small cross-sectional area in the conductor will be longer physically than an antenna having a larger cross-sectional area.

Very Short Antennas

The physical length of an antenna often poses a problem, especially in portable or mobile work in an automobile or boat. It is possible to use a very short antenna (at some sacrifice in overall efficiency) and electrically "load" it so it appears as an antenna of greater length. The missing portion of the antenna takes the form of a *loading coil* of wire wound on an insulated form (or supported by its own rigidity). The coil, for simplicity's sake, can be thought of as the missing antenna length, wound up into a compact structure. The coil may be placed at

CUBICAL QUAD ANTENNA is popular among CBers and radio amateurs alike. Quad beam delivers the most gain per area of any inexpensive beam antenna known. Build-your-own Quad data is in this Handbook.

one end of the antenna, or along the length of the antenna. You see these loading coils as bulges at the base of the antenna, or along its length. Coil design and construction is crucial, and the efficiency of the antenna depends upon the skill of design and manufacturing, and the expertise of the person making the antenna installation. It's not a game for beginners. In any event, the portion of the antenna used up in the coil contributes nothing to the radiation prowess of the antenna. Usually, the smaller the antenna in relation to the operating wavelength, the bigger the loading coil, the lower the antenna efficiency, and the weaker your signal! *There is no substitute for a full size, resonant antenna!*

Summary

The radio wave is an electric "undulation" which is a portion of the electromagnetic spectrum. The radio wave is specified in terms of size (wavelength) and frequency (cycles per second). It may be radiated and intercepted by an antenna which converts radio waves into high frequency electrical energy and vice-versa. The antenna has the same characteristics when transmitting or receiving and has the greatest efficiency when it is in a state of resonance.

Chapter 3

Your Antenna and Signal Interception

(How Does the Radio Wave Get Down That Skinny Cable?)

What happens when the radio wave meets the antenna? Why does a beam antenna make your signal louder? What is the meaning of antenna gain? Where does the gain come from? Good questions, and ones not easily answered.

To start with, a radio wave travels unhindered through space until it meets a conductor. In our case, the conductor is the antenna, and it is made of metal. By definition, a metallic object is one that has "free" electrons able to move about within the conductor. The intercepted radio wave imparts energy (motion) to these electrons, which move in a direction corresponding to the direction of the wave--along the conductor.

Inside The Antenna

The electron movement in the antenna element actually *reradiates a portion of the radio wave back into space* while the remainder is captured by the radio receiver attached to the antenna. This action takes place regardless of the length of the antenna and is shown in Figure 1.

When a radio receiver is attached to an antenna which is energized by a radio wave, a portion of power is extracted from the antenna and is passed to the receiver circuits. The total electron current flowing in the antenna (while very feeble) may be thought of as the sum of many individual currents, all acting along the length of the antenna. When all of the individual currents add up at the receiver, the maximum possible value of current is extracted from the antenna. This condition of maximum current is our old friend, antenna resonance. Resonance is established when the antenna bears a certain relationship to the length of the intercepted radio wave, usually found at multiples of one-half wavelength. The basic half-wavelength antenna is called a *dipole* antenna. In the CB band, a dipole antenna is about 18 feet long.

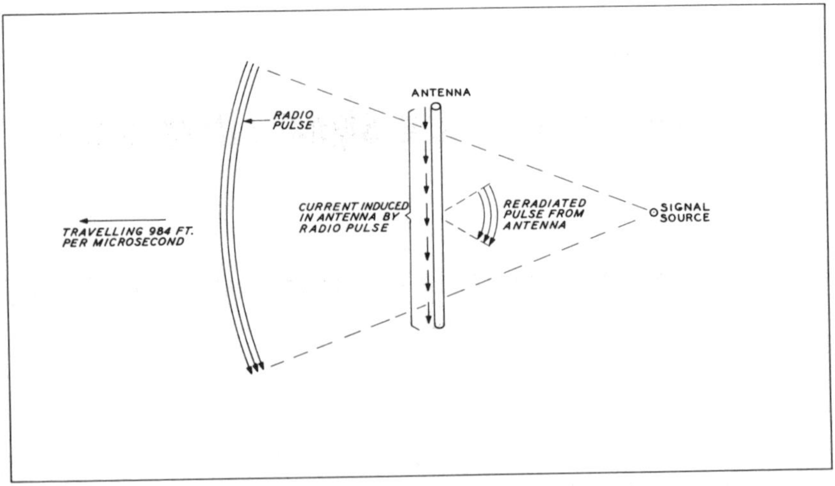

Fig. 1 RADIO WAVE MEETS ANTENNA. At the right is a radio signal source, such as your CB transmitter, radiating a radio wave moving to the left. For this example, the wave is just a short burst ("pulse") of energy. When the pulse meets the antenna (center), it causes free electrons to flow along the antenna in the form of an electric current. This current re-radiates a portion of the intercepted radio wave back into space. The original radio pulse passes to the left, beyond the antenna, followed shortly by the re-radiated energy from the antenna, which spreads out both to the right and left of the antenna. Now we have two pulses: the original one and a weaker, re-radiated one following along a fraction of a second later. When the current flowing in the antenna is a maximum value, the antenna is said to be "resonant".

Power Transfer

With any type of electrical power source, whether it be a storage battery, a generator or a radio transmitter, maximum power is delivered from the source to the destination *(load)* when the load and source each have equal voltage and equal current at their terminals. In this condition, the source *(radio transmitter)* and load *(antenna)* are said to be *matched*. Specifically, to obtain the maximum transfer of power from a radio transmitter to an antenna, or from an antenna to a receiver, the antenna must meet two requirements:

1- The antenna must be resonant at the frequency of the radio wave so as to deliver maximum signal current to the receiver, or to extract maximum power from the transmitter.

Figure 2

RADIO ANTENNA connected to load, which could be your receiver. When pulse meets parasitic, most of the radio energy is radiated back into space to loaded antenna. If phasing (timing) is correct, energy from parasitic element reinforces current flowing in loaded antenna. Yagi beams are made up of loaded antenna plus one or more parasitic elements.

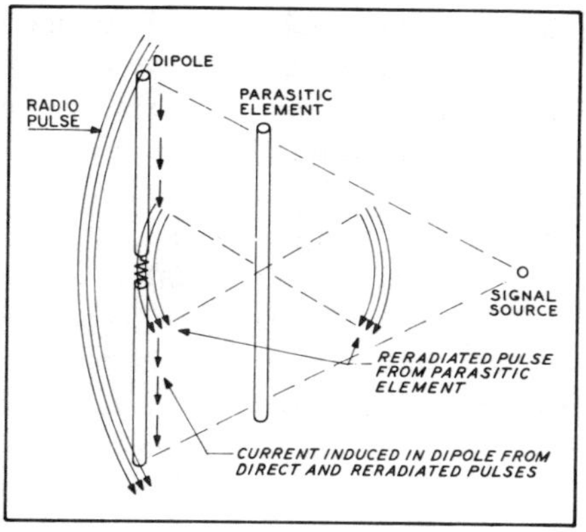

2- The signal voltage and current in the antenna and the receiver (or transmitter) must be matched to achieve best transfer of energy.

Even when these two demands are met, the matched, resonant antenna delivers only a portion of the current extracted from the radio wave to the receiver, reradiating the remaining portion of the current back into space a fraction of a second after the wave has arrived at the antenna. As you'll see, this reradiated wave can be made to perform in a very useful manner.

The Beam Antenna and Signal Gain

A *beam antenna* is an antenna that concentrates radio energy in a given direction at the expense of radiation in other directions. A good analogy is the headlight of an automobile which concentrates the light into a single beam, at the expense of illumination to the sides and back. A beam antenna concentrates the radio wave in a similar fashion, although the beam of an antenna is not nearly so well defined as that of a headlight beam.

A resonant antenna reradiates virtually all of the intercepted radio wave if it is not connected to a load. When such a device is placed near another antenna connected to a radio receiver, the unconnected antenna is called a *parasitic element*. Figure 2 shows a parasitic element placed near an antenna which is connected to a load. The parasitic element is placed between the antenna and the signal source. The antenna is now exposed to the direct radio wave from the source and *also to the reradiated radio wave from the parasitic element,* which reaches the antenna a fraction of a second later.

If the reradiated wave from the parasitic element reaches the antenna at the proper time, *it reinforces the radio wave received directly from the source by the antenna.* Adjusting the length of the parasitic element and its spacing from the antenna accomplishes this. Such a parasitic element is called a *director.*

When the parasitic element is placed *behind* the antenna and the signal source, and is adjusted to enhance the received signal, it is called a *reflector.* Beam antennas are made up of a connected, or "active" antenna, working in conjunction with a single reflector and one or more directors.

Antenna Power Gain

Power Gain (or signal gain) is a term used to express the power increase in receiving or transmitting of one antenna as compared to a a standard comparison antenna. The comparison antenna may be either a dipole, or it may be an imaginary device called an *isotropic antenna,* useful for mathematical computations in the laboratory. The isotropic antenna is considered to be so small that it radiates equally well in all directions--something that no actual antenna can accomplish.

Of great interest to the prospective purchaser or user of a beam antenna is the amount of power that still escapes from the sides and back of the array. The ratio of power radiated in the forward direction of the beam as compared to the amount radiated in the opposite (back) direction is called the *front-to-back ratio* of the antenna.

Beam antennas come in all shapes and sizes and prove to be a willing subject for imaginative manufacturers to exploit, sometimes at the expense of the buyer, dazzled by important sounding words and inflated claims. We'll go into beam antennas of various types in detail later in this Handbook. Before we do, however, it would be well to investigate and discuss the "yardstick" by which antennas of all kinds are measured. This "yardstick," when properly interpreted, can give you a lot of useful information about any antenna. The "yardstick" is called the *decibel.*

The Mysterious Decibel - The Unit of Truth!

Power gain and front-to-back ratio performance of beam antennas are expressed in terms of decibels (abbreviated *db*). The decibel is not a unit of power, but a *ratio of power levels.* In antenna practice, the decibel may be used as an absolute unit of measure by fixing an arbitrary level of reference. If the reference level is a dipole antenna

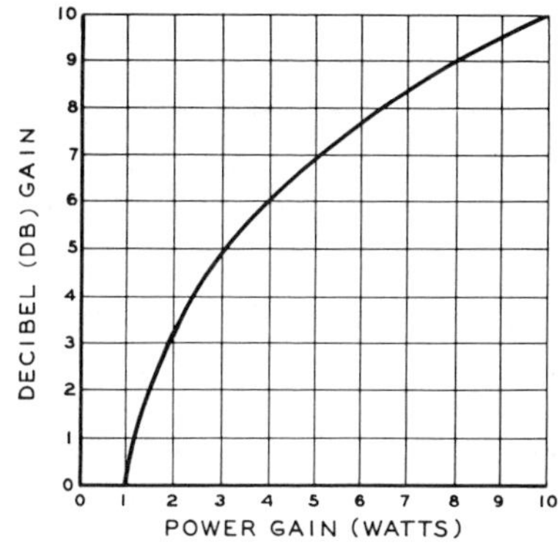

POWER GAIN OVER ONE WATT (UNITY)	DECIBEL GAIN
1.00	0.0
1.12	0.5
1.26	1.0
1.42	1.5
1.58	2.0
2.00	3.0
2.51	4.0
3.16	5.0
3.98	6.0
6.31	8.0
10.00	10.0

Fig. 3 THE DECIBEL--THE "YARDSTICK" OF PERFORMANCE. The decibel is the "yardstick" of performance in the electronics world just as "horsepower" is the yardstick in the automotive world. The decibel unit expresses the ratio between two power levels in an electrical circuit. Originally, the decibel measured the loss of voice power over one mile of telephone wire.

This chart shows the relationship between the decibel and the watt for gains of up to 10 over a reference level of one watt. For example, a power increase of 6.31 represents a decibel gain of 8. A decibel gain of 3 represents a power gain of 2.

Antenna gain is commonly expressed in decibels gain over a reference source, such as an isotropic radiator or a dipole. It is important to know the reference as the decibel is a ratio of change of power and can only be used as an absolute unit by fixing a level of reference. If a dipole is chosen as a reference antenna, it is said to have a power gain of unity (one).

(for example), another antenna may be said to have "so-many decibels power gain over the reference dipole."

Antenna gain expressed in decibels without mention of reference level is meaningless. It is often deliberately expressed in this fashion by some antenna manufacturers who hope to overawe the reader into thinking he is getting more power gain for his money than he actually is. For example, the statement, "Ten decibels power gain!!!" doesn't mean a thing, because it does not reference the gain to any level of measurement. If the statement said, "Ten decibels power gain over reference dipole" the statement has meaning, as the level of reference is given. More about this subterfuge later.

Shown in Figure 3 is a decibel chart and table giving the relationship between decibels and power. A power ratio of 4 provides a change of 6 decibels. A power ratio of 10 provides a change of 10 decibels. This relationship holds true regardless of the electrical circuit, and the decibel is commonly used in all electrical work throughout the world. It is a useful form of "shorthand" understood by all electrical engineers, regardless of whether they work with radio, audio circuits, power lines, or space communication. Like it or not, the decibel is here to stay.

A useful rule-of-thumb is that a power change of one decibel is *just noticeable* on a quiet communication circuit, and a power change of 3 decibels is about the smallest change that is economically feasible. That is to say, a power increase of less than 3 decibels is not worthwhile in most instances, as it is marginally helpful, and a power decrease of 3 decibels is not especially harmful.

Translated into terms of watts, this means that if your CB transceiver has a power *output* of 3 watts, boosting it to 4 watts by careful tuning won't be of much help! Or, if your power should drop from 3 watts to 2 watts, the fellow at the other end of the circuit probably won't notice the difference!

Doubling power provides a boost of 3 decibels, and that's about the least change that can really be noticed over a period of time. Accordingly, an antenna that provides 3 decibels power gain over a dipole could be worthwhile.

Experience has shown that a power increase of 5 decibels or so is definitely worthwhile, and a power change of 10 or more decibels is often startling in the results it achieves over a difficult communication path.

With this basic concept of decibels in mind, we'll investigate various antenna types in a later chapter, especially with respect to power gain. We'll build a "Truth Table" showing antenna power gain of various arrays compared to a standard antenna. But before we do this, let's turn to radio propagation and some common pitfalls along the road to better CB Communication.

Your Radio Range

The distance you can communicate on CB radio is called your *radio range*. It is measured along the surface of the earth which is neither smooth nor flat. As the radio wave travels along, it encounters various

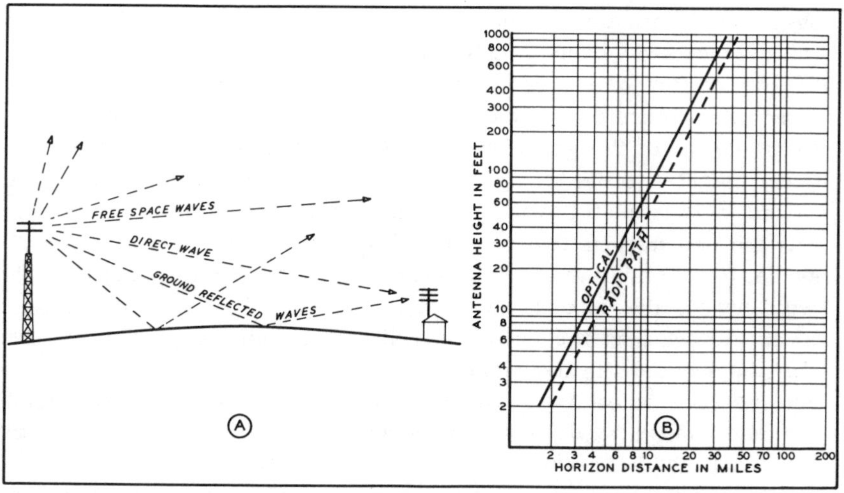

Fig. 4 YOUR RADIO RANGE. The CB radio wave may be thought of as a free space wave, a direct wave and a ground-reflected wave (A). The direct wave is useful over "line of sight" paths slightly greater than the optical path, as charted in drawing (B). As antenna height is increased, the radio horizon increases. Realistically, your ground wave radio range depends to a great extent upon placement of your antenna with respect to nearby metallic objects and wooded areas (see below). Erect your antenna clear of gutters, utility and telephone lines, stucco buildings, or other large structures. Trees often cast a "radio shadow" to a lesser extent. Hills may block radio signals and also tend to reflect the signal. TV pictures can be distorted and "ghost" images created by reflection of the TV wave from a hill or building.

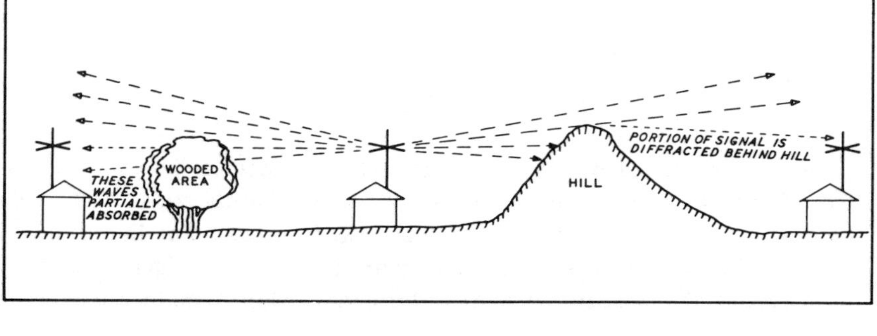

obstacles to its travel and gradually wanders off into space as the earth curves away beneath it (Figure 4).

The portion of the radio wave of interest to the CB operator is termed the *ground wave* which is the portion of the wave that hugs the earth. Range of the ground wave is determined principally by the antenna height and the topography of the land. Generally speaking, the ground wave is useful over *line of sight* distances. This is the distance between two stations that can "see" one another's antennas in a radio sense. Because the atmosphere exerts a slight bending effect upon the radio wave, the ground wave communication distance is about 1.3 times greater than the optical horizon. The radio range between two stations located "in the clear" is therefore the sum of the radio horizon distance of each station. Raising the antenna in the air extends the radio range to an appreciable extent, within the antenna height limits imposed by the Federal Communications Commission.

While this theory of radio range is nice, the practical communication range of CB radio stations varies greatly, especially in urban areas and cities having numbers of large buildings. With normal 27 MHz CB equipment and simple antennas, a base station-to-base station range of 20 miles or more is not uncommon. Beam antennas at both stations can often boost this range to 50 miles or more.

Base to mobile (car) communication range is usually less, normally about 5 to 10 miles over flat ground. In the city, surrounded by steel buildings, the range may be cut to a mile or less. Across water, the range of both base and mobile radios varies considerably and communication up to 30 or 40 miles is common.

Your radio path, whatever it is, is subject to interference over which you have no control. Other stations operating on the same channel can block an otherwise reliable radio path. Static, noise and interference can literally blot out a radio link at times. Good station equipment and an efficient beam antenna will permit the CB operator to make the best of his radio path and to obtain the maximum possible radio range. That's the name of the game.

Skip Signals

At certain times of the year, and under conditions beyond the control of the CBer, the radio range extends mysteriously out to thousands of miles, and CB stations across the United States boom in like local signals. These long distance signals are termed *skip signals*, as they skip across the country to distant listeners. Skip signals are created

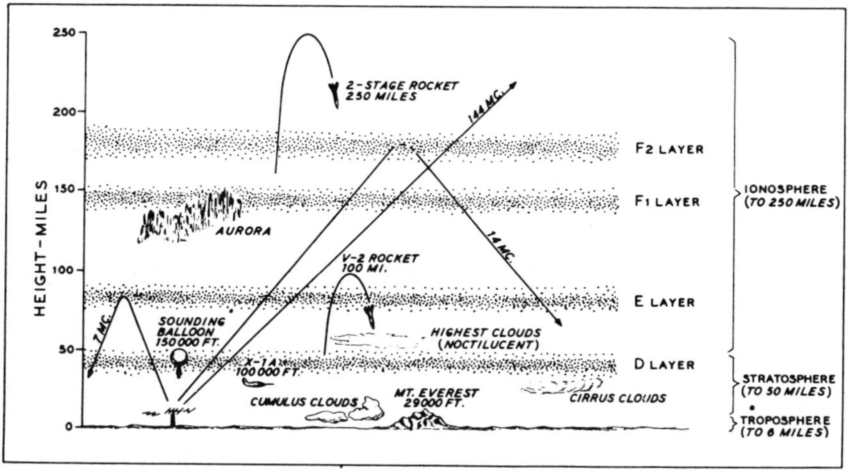

Fig. 5 THE RADIO MIRROR ABOVE THE EARTH. Layers of electrically charged air 60 to 250 miles above the earth reflect radio waves back to earth. CB waves are reflected by the F-1 and F-2 layers during daylight hours. At night, ionization of these layers is generally too weak to reflect signals, which then escape into outer space.

Layer ionization is a function of ultra-violet radiation from the sun, and varies from hour to hour and day to day. A long term variation (based upon sunspot activity) takes place over an 11-year cycle. During certain portions of the cycle, ionospheric ionization will not support skip signals on CB channels, while at other times, skip signals from thousands of miles away will be heard with amazing strength.

when the radio wave, instead of wandering off into space, is reflected back to earth at a distant point by the *ionosphere*. The ionosphere is an electrically charged portion of the atmosphere about 100 miles above the earth (Figure 5). The reflecting and refracting qualities of this gigantic "radio mirror" change from hour to hour, day to day and year to year. Generally speaking, the skip signals on CB channels are prevalent during fall, winter and spring months, and during daylight hours through the years of maximum energy radiation from the sun.

Radiation (other than light, such as ultra-violet and x-ray) from the sun waxes and wanes on an 11-year cycle. At the peak of the radiation cycle, the bombardment of the earth's atmosphere by the sun's energy creates an electrically charged layer of atmosphere, high above the earth, which reflects radio waves back to earth. The present period of high radiation activity started during 1977 and long distance CB skip interference has been very prevalent since then. The coming years will bring about intensified skip until about 1985, when the sun's radiation cycle will start to drop in intensity, building up once again as we go into the early "nineties".

In addition to skip signals caused by the action of the 11-year radiation cycle, quite frequently CB skip signals are noted in the summer months during early morning and late afternoon hours. The skip is "spotty," producing erratic signals from localized areas, up to 1200 miles away or so. This skip phenomena is called *sporadic-E layer* skip. It is caused by irregular "clouds" of electrically charged particles which reflect radio signals back to earth. These "clouds", seemingly quite small in size, have the ability to shift about very rapidly and to appear and disappear in a short space of time. Communication via sporadic-E "clouds" is common among radio amateurs operating on the higher frequency bands near 6 and 10 meters.

Other forces of nature are at work on the radio path, including ionospheric "storms" which blot out long distance radio communication for hours or days. While radio propagation has been studied for a number of years, and advance predictions of radio conditions are available from the National Bureau of Standards, radio transmission on the higher frequencies is still subject to unpredictable fluctuations. Although CB communications are supposed to be limited entirely to ground wave signals, signals from sky wave skip occur frequently and the CB operator should recognize and understand this event when it happens.

Truckers caught public's fancy in 1975 and explosion of CB radio brought a new era with problems in mobile communication. In mid-1977, FCC reported over 800,000 CB licenses issued monthly. As shown here, large trucks usually place CB antennas on side mirror mounts. Truckers are enthusiastic users of CB radio on the highways of America.

Chapter 4

The Curious Antenna

(Will It Ever Replace Smoke Signals?)

The concept of the antenna--a simple length of aluminum tubing or copper wire--taking electrical energy and transforming it into invisible radio waves that cross space at the speed of light is a breathtaking idea. Textbooks of great weight and complexity have been written about antennas, and very large antenna arrays have been built that defy comprehension by the layman. This Handbook ignores complexity and covers in detail the popular and everyday antennas encountered in CB radio and discusses their operation in simple, nontechnical terms.

You, the reader will not be placed in the position of the little girl sent to the encyclopedia to learn about penguins. When she finished reading, her mother asked if she was satisfied with what she had read. The girl replied, "Not really, the encyclopedia told me more about penguins than I wanted to know!"

This chapter (much shorter than an encyclopedia) covers properties of the antenna that are important to you. No complicated mathematics at all! Let's start with antenna polarization (Figure 1).

Antenna Polarization

Polarization of a radio wave is determined by the position and direction of the electric field with respect to the earth's surface. If the lines of the electric field of a radio wave are parallel to the ground, the wave is *horizontally polarized*. If the lines are perpendicular to the ground, the wave is *vertically* polarized. Since you can't see the field, and it takes instruments to determine the polarity, it's sufficient to know that, for the majority of CB antennas, the electric field is in the same plane as the antenna element so that a vertical antenna is ver-

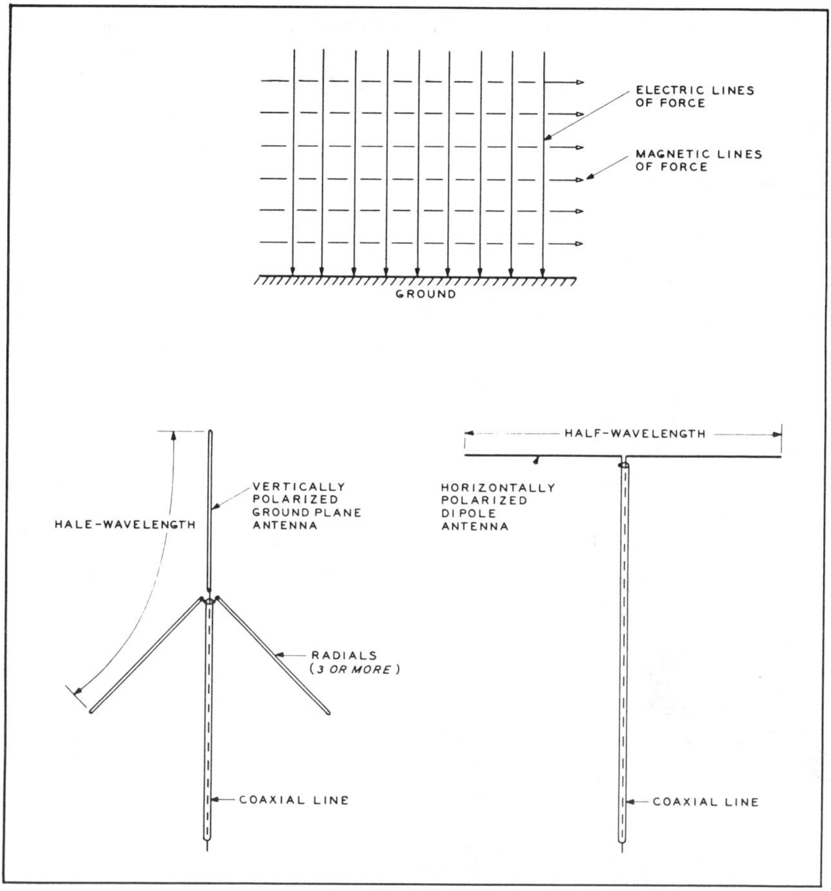

Fig. 1 CAN YOU SEE A RADIO WAVE? If you could, perhaps this is what you would see if a radio wave was travelling out of the page towards you! It is a fanciful drawing of a radio wave showing the electric and magnetic fields. Electric fields are commonly represented by solid lines and magnetic fields by dashed lines. The arrow heads indicate the direction of the field. In this illustration, the lines of the magnetic field are parallel to the ground and the lines of the electric field are vertical. According to the rules of the game, therefore, this wave is vertically polarized. For common antennas, the electric field is in the same plane as the antenna element so this represents the radio wave leaving a vertical antenna, such as a ground plane. Of course, the field surrounds the antenna, so this picture represents just a small portion of the wave. The magnetic field actually surrounds the antenna and the magnetic lines move outward, just as ripples in a pond spread when a stone is cast in the water. The electromagnetic field concept is a difficult one to grasp, so don't loose too much sleep over antenna polarization.

tically polarized and a horizontal antenna is horizontally polarized.

Antenna polarization is important in CB radio since optimum ground wave communication requires that the antenna at each end of the communication path have the *same polarization.* Cross-polarization reduces ground wave signals by 20 decibels or so! On the other hand, skip signals are relatively immune to cross-polarization, as reflection of a signal from the ionosphere tends to blur the polarization of the wave.

A degree of rejection of an unwanted signal may be obtained by reversing antenna polarization for a specific path between two stations. If both stations, say, are horizontally polarized, they can reject vertically polarized interfering signals by 20 decibels or so, thus improving their communication path.

Generally speaking, CB antennas are placed in a vertical position and the signals are vertically polarized. This was originally done so that base stations could have good communication with mobile stations using whip antennas which are vertically polarized.

Some CBers, in order to reduce local interference, have shifted over to horizontal polarization. Experiments have shown that horizontally polarized antennas reduce certain types of electrical interference (auto ignition noise, for example) while still producing good signals over a line-of-sight path. The great majority of CBers, however, prefer vertical polarization. Beam antennas may be mounted in either a vertical or horizontal position, at the choice of the user, and some provide switchable polarization controlled from the operating position. This is a distinct advantage.

Antenna Operation

While you can't do very much about the tricks the ionosphere may play on you, there's plenty you can do about having a good, efficient antenna system. Let's turn, therefore, from radio propagation to the antenna itself and see what goes on inside this interesting device. The remainder of this chapter discusses some of the "ground rules" of antenna operation.

Impedance --- More Power To You!

You'll often see terms such as *surge impedance, feed impedance, radiation resistance, input impedance* and the like, used with reference to antennas. These terms are telling you something you should know about the antenna. The big question is: what's it all about?

The simplest answer is that when the complete antenna system is in a state of electrical balance (the state of balance being defined by these various terms), the overall operation of the system is at the highest possible efficiency. This means the best reception and the strongest signals.

Many readers of this Handbook are familiar with Ohm's Law, which is:

$$R = \frac{E}{I}$$

Simply, this formula tells you that the resistance of any electrical circuit (R) is proportional to the *ratio* of the voltage (E) to the current (I) in the circuit. This is *not* the definition of Ohm's Law in most books, which usually ends up talking about resistance in terms of ohms, and in terms of physical resistors. As far as antennas go, the idea of *resistance*, as such, is misleading, and it is much better to think of the word "resistance" as expressing the ratio of the voltage to the current at any given point in the antenna circuit. Thus at a high "resistance" point, the ratio of voltage to current is high and at a low "resistance" point, the ratio of voltage to current is low.

Remember in Chapter I the short discussion of Maxwell's discovery of the interplay of energy between the electric and magnetic fields *around* a wire, and the current flowing *within* the wire? Ohm's Law, in the broadest sense applies in this case, describing the ratio between the electric field and the current flow in terms of a fictitious resistance, which may be termed the *radiation resistance* of the wire, or antenna:

$$\text{Radiation Resistance of the Antenna} = \frac{\text{Voltage field about the antenna (E)}}{\text{Current flowing within the antenna (I)}}$$

or, in Mr. Ohm's terms: $R = \frac{E}{I}$, which is the same law that applies to direct current circuits, car batteries, your refrigerator motor, the TV set, and such.

Radiation Resistance

To repeat: the terms radiation resistance, input impedance, surge impedance, feed impedance, etc., all refer to the *ratio* between the voltage field about a conductor and the current flowing within the conductor. The expression of this ratio is in terms of ohms. This term is, perhaps, unfortunate, as it implies some kind of actual resistance within the antenna, which does not really exist. As a dodge, this "resistance" is referred to as radiation resistance, which is OK as long as you understand that it means a *ratio of voltage to current*, and

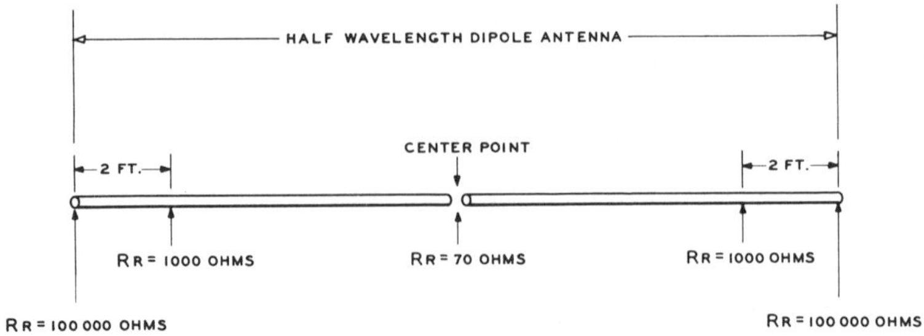

Fig. 2 WHERE DOES THE POWER IN AN ANTENNA GO? It has to go someplace, and it does. It is radiated into space in the same manner as heat and light. It leaves the antenna, never to return. It is difficult to conceive of radio power vanishing into empty space, so a fictituous term (radiation resistance) is used to describe the radiating effectiveness of the antenna. Radiation resistance, as explained in the text, is an expression of the ratio between the voltage field of the antenna and the current in the antenna at any given point along the antenna. In a simple dipole, the radiation resistance at the center is quite low, increasing towards the ends of the antenna. Thus, the voltage to current ratio is different at different points along the antenna. Most antennas are fed at the center and--in a dipole--this is the point of lowest radiation resistance. The dipole in this illustration is suspended in space about a half-wavelength above the ground, and the measurements shown may actually be found when the proper instruments are used. Moving the antenna closer to, or away from the ground, will alter all the measurements drastically.

not an actual value of resistance.

Let's take an example: Suppose 3 watts of radio frequency (r-f) power is applied to the center of a dipole antenna. With appropriate (and expensive) instruments, it is determined that the current flowing in the dipole at this point is 0.21 ampere and the voltage field about the dipole is 14.7 volts (Figure 2). The radiation resistance (R_R) is, therefore:

$$\text{Radiation Resistance} = \frac{E}{I} = \frac{14.7}{0.21} = 70 \text{ ohms}$$

No 70 ohm resistance exists in the antenna. But, if the antenna were removed, and a 70 ohm resistor substituted in its place, the *same ratio* of voltage to current would exist, and the same absolute values of voltage and current.

Now let's move out along the dipole's length and measure the voltage and current towards one end, say about two feet from the end of a CB dipole. What do we find? The same amount of power exists at this point (3 watts) but the voltage field has increased to 55 volts and the current in the dipole has dropped to .055 amperes. The fictitious radiation resistance value is now expressed by:

$$\text{Radiation Resistance} = \frac{E}{I} = \frac{55}{.055} = 1000 \text{ ohms}$$

Finally, let's examine conditions at the very tip of the antenna. Logic tells us that current is zero (since it has no place to go) and voltage should therefore be infinite, if Ohm's Law is obeyed. Practically, this is not true, as the current can "leak" into the atmosphere, or into the insulators that support the dipole. The same 3 watts exist, but the instruments show that the voltage field is now 550 volts and the current is .0055 amperes. The radiation resistance, accordingly, is:

$$\text{Radiation Resistance} = \frac{E}{I} = \frac{550}{.0055} = 100{,}000 \text{ ohms}$$

So! While the power at any point in the antenna is constant, the voltage to current ratio varies all along the antenna, and the radiation resistance (or whatever you want to call this ratio) varies too. The fact of interest to the antenna designer is the ratio *at the point at which he feeds energy to the antenna* by attaching a transmission line. This is usually at the center point, since the radiation resistance ratio is reasonably low and manageable. If the radiation resistance is too low or too high, other problems arise in feeding energy to the antenna which are beyond the scope of this Handbook. Suffice to say that radiation resistance values of 25 to 100 ohms or so are a "target" range for antenna feed points.

Well, What Does It All Mean?

In summary, then, the various terms mentioned previously expressing "radiation resistance" are *not* expressions of actual resistance itself, but an expression of a ratio between the voltage field of the antenna and the current within the antenna at any given point along the antenna. If you can forget the traditional concept of resistance and think instead of the *ratio of voltage to current,* you can more readily understand some of the more obscure representations of working antennas and the over-complex and confusing terms that express these concepts.

Impedance And All That!

So far we've looked at radiation resistance. What about impedance, another OK-term bandied about by the knowledgeable to impress the natives? Remember---the term *resistance* is used in antenna work in its broader interpretation as the voltage to current ratio at which power is consumed or transferred. The antenna, of course, does not consume power, it merely transfers it into the space about it. Nevertheless, for the purpose of circuit design it is convenient to substitute the resistance concept because the circuit *behavior* conforms to that of an actual, real-life resistance.

The term *impedance* is used in a comparable sense, too. It is also a voltage to current ratio, but it expresses an even more general concept than resistance because it implies that all of the power supplied to the circuit *may not be consumed or passed on*, but a certain proportion of it can be returned to the source during some part of the radio energy cycle. The proportion of power consumed or passed on is expressed in terms of *radiation resistance* and the portion of the power returned to the source is expressed in terms of *reactance*. Reactance is undesirable because it makes the antenna difficult to "load" (take power from) the transmitter.

Purists draw a distinct difference between resistance and reactance; however, in antenna talk, the terms are often interchanged. This can be done in the case of a resonant antenna without stretching the truth. And in the case of a CB antenna, operating on various closely spaced channels, the terms can be interchanged readily, even though there is a distinction between radiation resistance and impedance or reactance that will be more easily understood when we tackle the problem of transmission lines and standing wave ratio later in this Handbook.

Now, we have a vague idea of radiation resistance and the various terms that mean the same thing, and we know that the voltage and current ratio is different at different points along the antenna. A very nice picture can now be drawn of the variations, as explained in the next section.

Antenna Voltage and Current

Now look at the drawing of Figure 3. This shows two antennas, one a full wavelength antenna and the other a half-wavelength antenna. The "radio wave" can be drawn on these figures and now, based upon the previous discussion, it can be seen that this is a current wave, hav-

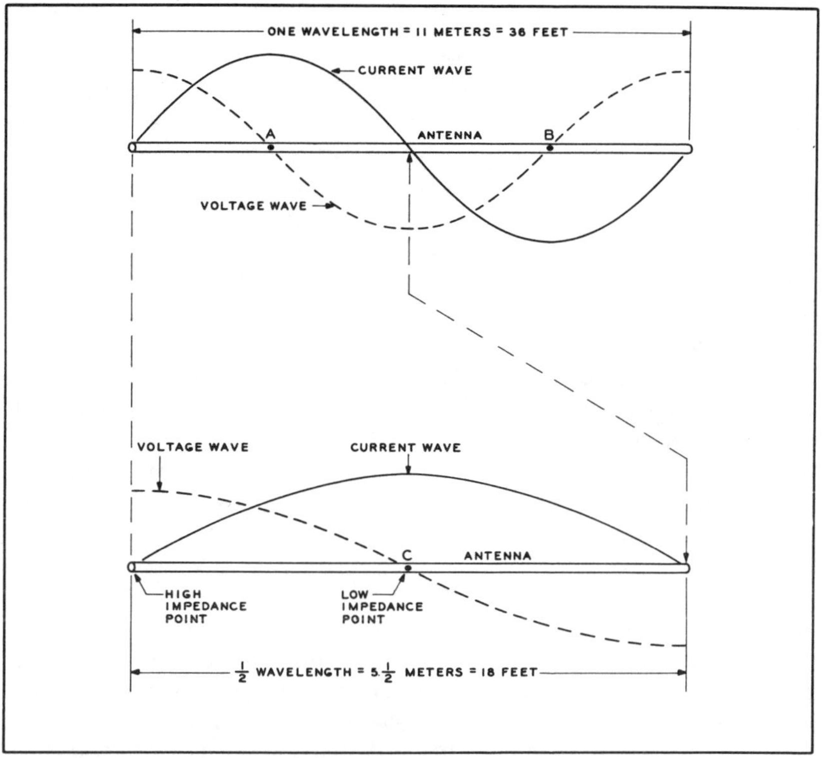

Fig. 3 DID YOU KNOW YOUR ANTENNA LOOKS LIKE THIS? Shown at the top is an antenna one wavelength long for the CB channels. At the bottom is a half-wavelength antenna. The voltage and current waves for a radio signal are drawn on the antennas. The height of the wave indicates the strength ("amplitude") at that point on the antenna. The voltage wave (dashed line) is maximum at the high impedance points on both antennas, and is a minimum at the low impedance points. The current wave (solid line) is exactly opposite, being a minimum at the ends of the antenna (the high impedance points) and a maximum at the low impedance points. In the case of the half-wavelength antenna, this is at the center.

You don't believe it? Well, if sufficient radio power was applied to either antenna (100 watts, or so), the voltage wave could be traced out by holding a neon lamp or flourescent tube near the antenna. The lamp will glow brightly from induced power at the high impedance points and will be extinguished at low impedance points. The lamp need not touch the antenna for this experiment.

For convenience, CB antennas are usually fed at low impedance points, as low impedance coaxial transmission line is readily available at modest cost. In the case of the one wavelength antenna, the feedline would be attached at either point A or point B. The feedline is attached at point C for the half-wavelength antenna.

ing a minimum value at the ends of the antenna and a maximum value--in the case of the half-wave antenna--at the middle. Also observe that the one wavelength antenna has two current maximums on it.

Figure 3 can now be modified to show the voltage wave as well as the current wave for both antennas. The voltage wave is a dashed line. The voltage wave is maximum at the high resistance ratio (impedance) ends of the antenna and the voltage curve is exactly opposite to the current curve. That is, the crest of one falls at the low point of the other. Radio lingo says these waves are "out of phase". Note, too, that *voltage is always maximum at the end of the antenna*, and a minimum one-quarter wavelength along the antenna from the end.
Likewise, *current is always minimum at the end of the antenna* and a maximum one-quarter wavelength along the antenna from the end. Finally, it should be noted that energy is usually (not always) introduced into the antenna at a point of maximum current (at the crest of the current wave). Thus, the dipole antenna is fed at the center--- the point of highest current.

Upwards And Onwards

Now that you have an idea of some of the terms and concepts used in antenna work, and a bit of the interesting history behind the development of today's antennas, let's look at some specific CB antennas. Antennas are like blondes:--"There are only two types of blondes---good and better." So, let's look at some of the better ones (antennas, that is).

ANTENNAS ARE LIKE BLONDES--there are only two types: good and better!

Chapter 5

Everything You've Always Wanted To Know About CB Antennas

(But Were Afraid To Ask!)

Man! How many types of CB antennas exist?

You can purchase a bewildering variety of exotic looking antennas for sums up to hundreds of dollars, or you can build some of the better ones yourself from data given in this Handbook.

Which antennas are the good ones, and which are the "lemons"? All too often advertisements are seen extolling the virtues of some new form of antenna. Can you believe what you read?

There are no easy answers to these questions, but there are guidelines in this Handbook which, when properly applied, can help you buy or build the proper antenna for your circumstances. These guidelines also answer many of the questions surrounding antenna operation.

What Makes A *Good* CB Antenna?

The satisfactory CB antenna is *one that works* and *is preadjusted,* or has a noncomplicated adjustment that affords proper tuning. The antenna should be easily put together and erected, should not fall apart in a windstorm, and should be rust- and corrosion-proof. A big order, but all of the antennas described herewith (including most of the good ones you can purchase) meet these requirements. Once they are properly assembled and placed in position, they are ready to be used. Above all, the antenna should be *honest*. That is, the operation should live up to the advertised virtues of the antenna.

In case of a beam antenna, in addition to ease of assembly the antenna should *work like a beam.* That is, it should make the transmitted and received signals louder in the chosen direction and should reject signals coming in "off the beam" (from unwanted directions).

It's possible to determine the excellence of an antenna, both from the study of the configuration of the antenna and also from actual on-the-air tests. Both techniques are discussed at length in this Handbook.

To aid you in understanding antenna operation, a unique "Truth Table" is given in Chapter 8 covering various types of antennas. The table provides you with a simple and accurate standard of comparison by which you can judge antennas. Armed with this information, you can evaluate the more popular antenna types and see what makes them work, how they work, and estimate their probable power gain. Re-read the section of Chapter 3 on the decibel as a "yardstick" before you jump into the next sections!

The Dipole Antenna

The basic shortwave transmitting and receiving antenna is the *half-wave dipole* shown in Figure 1A. Its length is equal to about one-half the transmitting wavelength, and it is usually made of wire, supported at the ends by insulators. The dipole is broken at the center, at which point a two-conductor feedline is attached. The dipole can be mounted in any position, but it is usually mounted horizontally or vertically between two supports. The length of a wire dipole for any shortwave frequency may be computed from:

$$\text{Length (feet)} = \frac{468}{\text{Frequency (MHz)}}$$

For example, the length of a half-wave dipole at 27 MHz (near the mid-point of the CB channels) is:

$$\text{Length} = 468/27 = 17.3 \text{ feet, tip to tip}$$

When erecting the dipole antenna, it is important to make sure that the two-conductor feedline does not interfere with proper antenna operation. To accomplish this, the line is led away at right angles to the antenna, as shown, when the antenna is in a horizontal position. When the dipole is mounted vertically, and it is desired to bring the feedline away below the antenna, a different scheme is used. The bottom of the

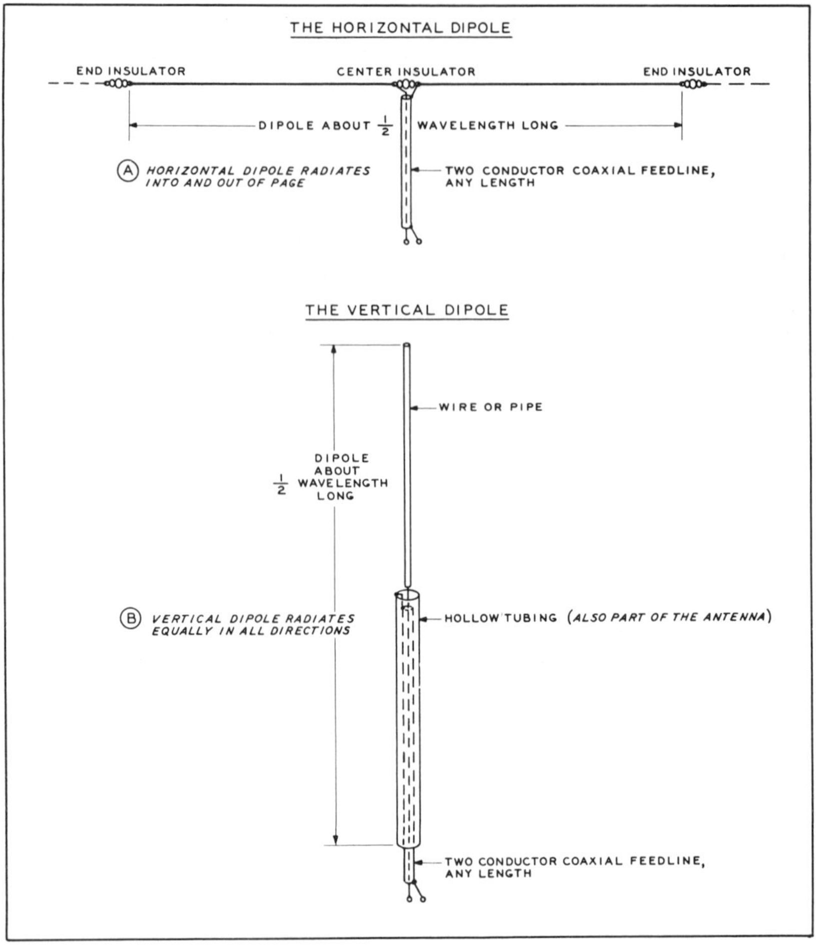

Fig. 1 THE DIPOLE ANTENNA is the fundamental shortwave antenna for both transmitting and receiving. Mounted horizontally (A), the dipole is commonly made of wire, supported at the ends by insulators. The low impedance center point is broken and a two conductor coaxial feedline is attached to the halves of the antenna. The feedline should be brought away at right angles to the antenna since it may have a radio field about it which can distort the radiation pattern of the antenna. This dipole radiates maximum energy into and out of the page.

The dipole antenna also may be mounted vertically (B). In order to bring the two conductor coaxial feedline away from the antenna without disturbing the radiation pattern of the antenna, the bottom half-section of the dipole is made of hollow tubing, with the insulated feedline running down the inside. Since the radio energy is concentrated on the outside of the dipole, the inside of the bottom section may be used as a shield for the feedline.

dipole section can be made of hollow metal tubing instead of wire, and the insulated feedline is passed down the center of the bottom section as shown in Figure 1B. Since the radio energy is concentrated on the *outside* surface of the dipole, the *inside* of one section may be used as a shield for the transmission line, provided the line does not touch the tubing wall. This effectively isolates the feedline from the electric field about the dipole. The dipole radiates radio energy at right angles to the direction of the conductor, as shown in the illustrations. The horizontal dipole thus has a bidirectional "figure-8" pattern and the vertical dipole has a nondirectional, circular pattern. As in many instances, the picture is worth a thousand words in this case.

Before we leave the dipole for the moment, it is important to note that this simple antenna is often considered to be the "standard" of comparison against which the performance of other antennas may be measured. There are other standards of comparison, and we'll discuss them when the time comes. But for now, remember the dipole as the comparison antenna --- the principal "yardstick" for the "Truth Table" in Chapter 8.

The General Coverage Antenna – The Ground Plane

The simplest, most popular and least expensive CB antenna is the *general coverage antenna*. This device radiates, or sprays, radio energy in all directions (except straight up). A typical general coverage antenna is the *ground plane*, shown in Figure 2. The ground plane consists of a vertical antenna, or *radiator*, mounted above several horizontal rods, or *radials*. The radiator and radials are all usually about a quarter-wavelength long, or nearly 9 feet for CB frequencies. The length of a quarter-wave element is:

$$\text{Length (feet)} = \frac{234}{\text{Frequency (Mhz)}}$$

The vertical section is considered to be the antenna proper, and the radials establish an artificial ground, or ground plane (hence the name) at the base of the radiator. Looking at it another way, the ground plane antenna is a vertical half-wave dipole, fed from below, with the bottom half of the dipole split into separate radials which are swung up into the horizontal plane. The antenna is fed with a two-conductor transmission line.

A general coverage antenna can be effectively used if it is desired to communicate with stations located in many different directions about your CB station. If, for example, you are conversing with a mobile

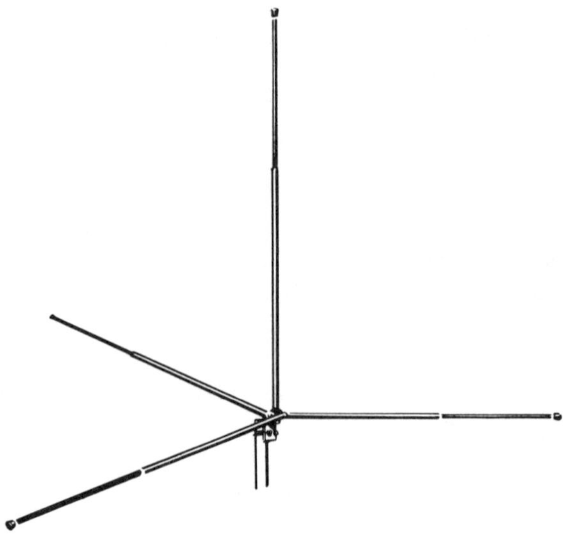

Fig. 2 THE POPULAR GROUNDPLANE. This inexpensive CB antenna is a close relative of the vertical dipole shown in Figure 1. It has a vertically polarized field and a nondirectional pattern, "spraying" radio energy in all directions about it (except straight up and down). The vertical rod is the antenna, and is a quarter-wavelength long -- about 9 feet for the CB channels. The three horizontal rods are called "radials" and form an artificial ground ("ground plane") directly beneath the antenna.

The ground plane antenna is derived from the vertical dipole, with the bottom half of the dipole split into separate radials which are swung up into the horizontal plane. Some ground plane antennas have drooping radials which modify the electrical characteristics of the antenna so that it more closely matches a 50 ohm coaxial transmission line.

The radials may be connected directly to the supporting mast, as shown here, but the antenna portion of the ground plane must be insulated from radials and mast. The center conductor of the coaxial feedline is attached to the antenna, with the shield connected to the radials.

station that moves about the countryside, a ground plane antenna at your base station insures that the mobile always remains within the radiation pattern of your antenna as long as it is within radio range of your station. Signals from a general coverage antenna of this type spread out much as the concentric circles spread from a stone cast into a quiet pond.

On the other hand, since a general coverage antenna receives and transmits in all directions, it does not provide the user with any protection from interfering signals on the channel in use, since all signals are received regardless of the direction of the incoming signal.

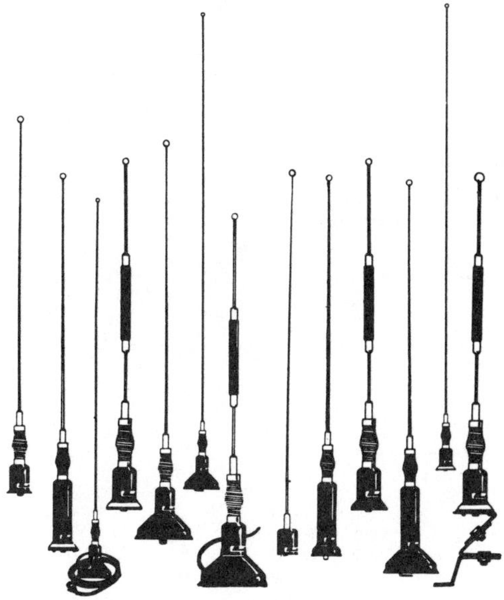

Fig. 3 WHIPS, WHIPS, WHIPS! Mobile whip antennas come in all shapes and sizes. Shown at right are full length and coil-loaded whip antennas for mobile service. Mini-whips are base or center loaded. These antennas have special mounts so user can place whip at the most advantageous spot on car.

A variation of the ground plane is the whip antenna (Figure 3), which is mainly used for mobile work. The whip usually takes the form of a stainless steel, tapered rod and is often seen on police cars, the autos of mobile radio hams and on CBer's automobiles. A full-size CB whip is a quarter-wavelength long, or nearly 9 feet. *Coil loaded* mini-whips may be purchased for CB service that are as short as 3 feet, the difference in length being made up in a loading coil placed at the base or center of the whip. Some varieties of whip antennas are made of Fiberglas for greater strength and flexibility, with a copper strap buried within the tough, outer insulation. The whip differs from the ground plane in that radials are not used, the body of the automobile or carrier serving as an artificial ground system. More about this in a later chapter.

The Extended Vertical Antenna

Vertical antennas are not limited to one-quarter wavelength in length. In fact, the half-wave and five-eighths wavelength vertical antennas provide somewhat improved results over the more common ground plane. These antenna types are shown in Figure 4.

The half-wave vertical antenna is shown in illustration A. The dipole ends are points of high voltage and high impedance and the

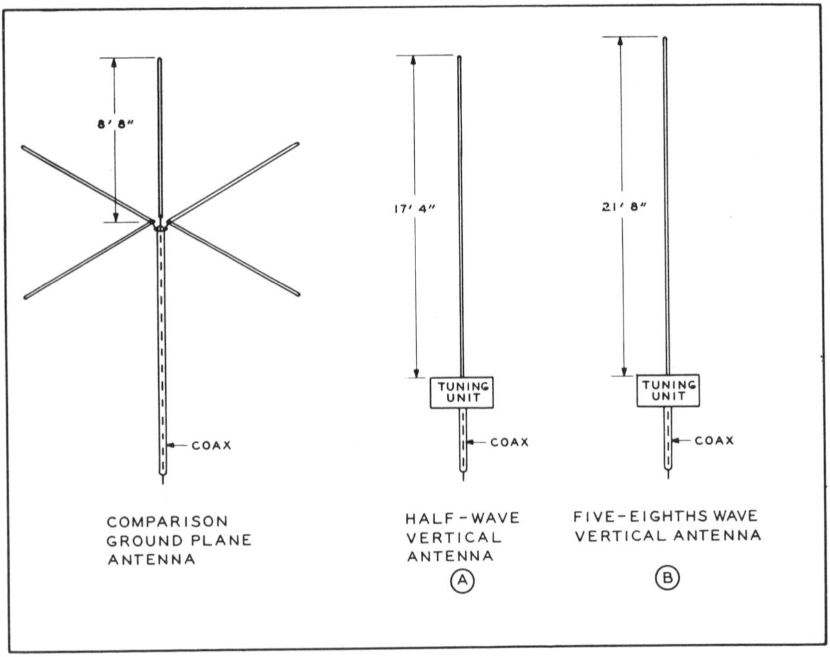

Fig. 4 EXTENDED WHIP ANTENNAS provide boost in gain over popular ground plane antenna (left). The half-wave vertical antenna (A) is about 17½ feet high and provides boost in power gain of 1.8 decibel over ground plane. The five-eighths wave vertical antenna (B) is about 21½ feet high and provides 3 decibel boost in power over ground plane. This is a worthwhile increase in signal and an antenna of this type is recommended for general coverage work. The main disadvantage of any vertical antenna is that it is very responsive to automobile ignition noise which is vertically polarized.

center point is at a relatively low voltage and low impedance. Some kind of matching device must be included to match this impedance to that of the transmission line. This device, or network, is mounted at the base of the antenna and may be included as part of the antenna itself, as described in a later chapter.

The half-wave vertical antenna is a complete dipole section and requires no radials in most installations. This is a great advantage where space is at a premium. The half-wave vertical antenna provides a power gain of about 1.8 decibel over the quarter-wave ground plane, a worthwhile increase in gain for such a simple antenna. The pattern is nondirectional, and general coverage is provided.

Fig. 5 HALF-WAVE VERTICAL ANTENNA has peculiar "ground plane" device at the bottom. This manufactured antenna has novel tuning network at the base that resembles a coiled ground plane. The inductor is a single turn coil about one foot in diameter which is connected in parallel with a fixed capacitor. The capacitor is an integral part of the mounting base and is formed by the end of the antenna itself which passes into a plastic sleeve inside the antenna mount. The capacitance of antenna to mount forms the fixed network capacitor. Coil and capacitor are adjusted to resonate at 27 MHz. The coaxial transmission line is tapped onto the one turn coil at the proper matching point. Adjustable tap is seen at right side of base coil. A very clever assembly!

The five-eighths wavelength vertical is shown in illustration B. The extra length of the radiator provides additional power gain (a gain of 3.0 decibels over a quarter-wavelength ground plane) and also eliminates the need for the horizontal radials. The five-eighths wavelength vertical antenna is popular as a commercial broadcast antenna as it provides the greatest possible gain for a simple vertical antenna while still maintaining an omnidirectional pattern. It is also extensively used for industrial and police mobile service in the 27 MHz and VHF regions.

The Directional Antenna

The *directional antenna* radiates and receives a maximum of energy in a *wanted* direction at the expense of radiation and reception in other directions. Most all antennas exhibit some degree of directivity, even the dipole, which radiates most of its energy at right angles to the

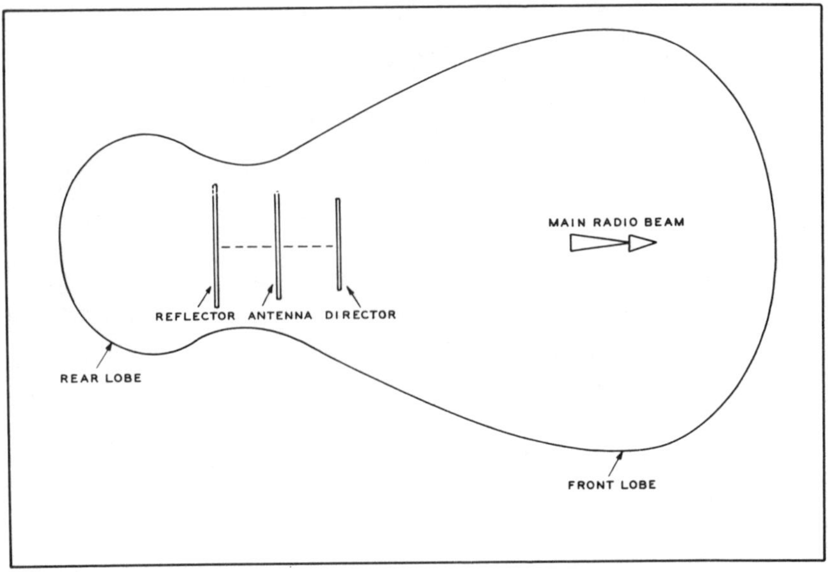

Fig. 6 DIRECTIONAL YAGI BEAM ANTENNA transmits and receives r-f power in one direction at expense of transmission and reception in other directions. A typical vertical CB beam antenna has a radiation pattern somewhat as shown above. This view is looking at the plane of the elements from the side of the beam. In order to "tighten" or "squeeze" the pattern to achieve a narrower beam, a larger antenna having more parasitic elements is required. Note that the beam antenna still has a small radiation lobe to the rear, and some radiation off the sides of the array.

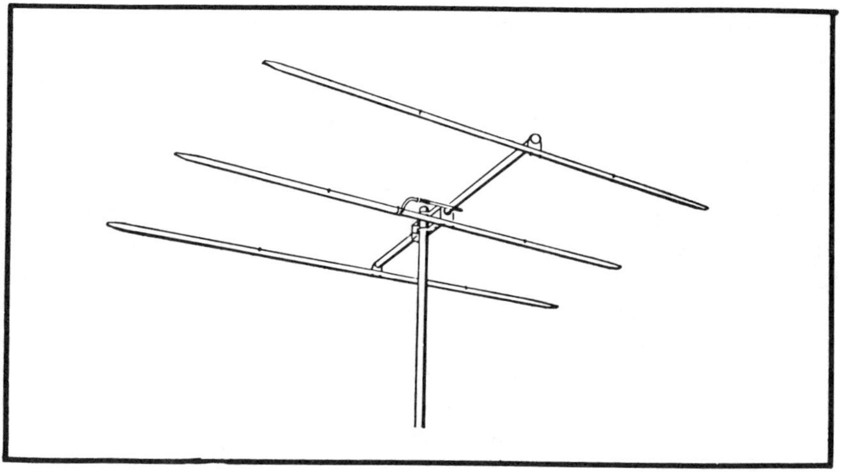

Fig. 7 HORIZONTALLY POLARIZED BEAM ANTENNA. Most CB antennas are vertically polarized, but the Yagi beam may be easily "flopped" over to horizontal polarization, as shown here. Horizontal polarization reduces interference from vertically polarized signals to some extent and also reduces ignition interference from automobiles and trucks, which is usually vertically polarized. Ionosphere-reflected signals tend to loose polarization.

wire. If the antenna is designed for maximum directivity, it is called a *beam antenna*. The two most popular beam antennas are the *Yagi* and the *Quad* and these will be discussed in the next chapter. These antenna types concentrate the radio wave into a single, broad lobe which may be aimed in the desired direction, much like a flashlight or the headlight on a car. By proper design, the energy radiated in other directions is reduced. This "signal amplification" is of great benefit to the user, and applies equally well when the antenna is used for either transmission or reception. In many cases, the use of a beam antenna at both ends of a difficult communication path permits reliable communication that otherwise would be very difficult if simple, general coverage antennas were used. The concept of a radio beam is shown in Figure 6.

Vertical or Horizontal Polarization?

The beam antenna may be mounted so as to be either horizontally or vertically polarized. Most radio amateurs use horizontal polarization from custom; however, CBers seem to tend towards vertical polarization. Generally speaking, vertical polarization should be used when communicating with a mobile or fixed station using a vertical antenna. Horizontal polarization should be used between stations wishing to reduce interference from vertically polarized stations on the same channel. For line-of-sight distances, either polarization is equally effective, although vertical polarization tends to pick up ignition interference from passing vehicles more readily than does horizontal polarization.

Within groundwave range, working from one polarization to another will drop signals as much as 20 decibels (over three S-units). Horizontal polarization will reduce ignition interference (which seems to be vertically polarized) by about the same degree. Skip signals are relatively immune to polarization, since reflection of a radio wave from the ionosphere results in random polarization of the received wave. A choice of polarization is nice to have at your fingertips, and on occasion may "save" your communication circuit under conditions of heavy interference.

The "Wonder" Antenna--You Wonder How It Works!

Every so often some joker comes along with a brand new antenna design that he extolls as the greatest thing since sliced bread. It provides (so you are informed) more gain, or is smaller in size than other antennas, and is the answer to every CB operator's prayers.

Beware! Nature's laws have not been repealed and you cannot get something for nothing. Be wiser, not sadder! Read the next chapter, and you can judge the "Wonder" Antenna yourself!

Chapter 6

More Antenna Gain For You!

Part I. The Yagi Beam Antenna

The classic beam antenna is the *Yagi*, or *parasitic beam*, named after Dr. Hidetsugu Yagi of Tohoku University, Japan. Dr. Yagi built the first simple, high gain antennas named after him in 1926 and beams of this type were popularized by radio amateurs just before World War II. They were extensively used for radar antennas during the war and are now used for high frequency and VHF work where low cost and high gain per unit of size are demanded. Yagi antennas come in various sizes, and a summary of the more popular versions follows.

The parasitic antenna element was briefly discussed in Chapter 3. A Yagi beam is made up of a dipole antenna and one or more parasitic elements in the same plane which may function as a reflector or director. Generally speaking, the greater the number of parasitic elements, the greater the power gain of the Yagi antenna. As the gain increases, the sharpness of the beam increases, too (Figure 1).

The Yagi for CB Radio

Yagi CB antennas are usually constructed of aluminum tubing and the active dipole and parasitic elements are approximately a half-wavelength long, or about 18 feet. Proper antenna operation depends not only upon element length, but also upon the spacing between the elements, which usually runs about 0.2 wavelength, or around 7 feet for the CB channels. Element spacing and lengths will vary, of course, depending upon antenna design, but the above figures are about "par".

The simplest Yagi is the *two element beam*, made up of a dipole and one parasitic element. The parasitic may be placed behind the dipole (a reflector element) or it may be placed in front of the dipole (a director element). Whether the parasitic element works as a reflector or a director depends upon the length of the element and the spacing from the dipole. The two element Yagi beam has a gain over a reference dipole of about 5.0 decibels. This is equivalent to a power gain of about 3.2, making the 5 watt power input of your CB station equivalent

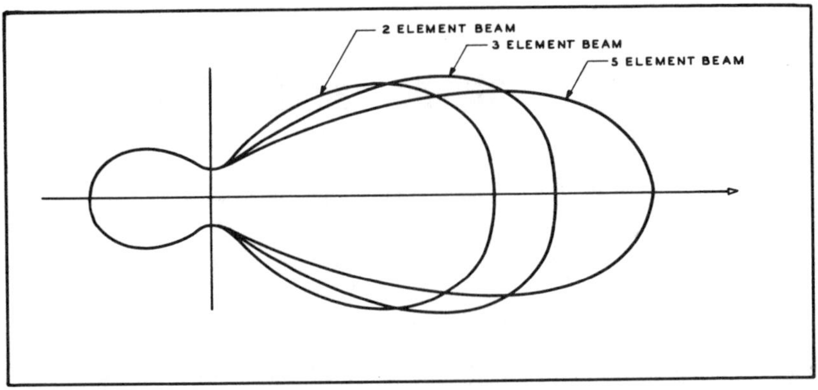

Fig. 1 BIG BEAMS GIVE NARROW PATTERNS. As parasitic elements are added to the beam antenna the pattern becomes narrower. An approximation of added gain is one decibel per parasitic element after the first reflector and director. Four element Yagi has a power gain of about 10 decibels and five element Yagi will have a power gain of about 11 decibels over a dipole. Adjustment of beam antenna having more than five or six elements is complicated and time consuming. Sharp beam pattern requires accurate aiming of antenna for maximum signal response.

to 16.0 watts. This is certainly a worthwhile increase in signal power and it is useful in receiving as well as in transmitting.

Next larger in complexity and size is the three element Yagi beam. This array has a reflector and a director in addition to the dipole, and is a very popular CB antenna. The power gain of the three element Yagi is about 8.0 decibels over a dipole antenna, equivalent to a power gain of 6.3. This is equivalent to running about 31 watts input to your CB station -- and that ain't hay! And it is perfectly legal, too!

More parasitic elements may be added as directors to the Yagi array and an approximation of gain is the addition of one decibel per parasitic element after the first two elements (Figure 2). That is, a four element Yagi (dipole plus three parasitic elements) will have a power gain of about 10 decibels, and a five element Yagi (dipole plus four parasitics) will have a power gain of about 11 decibels over a comparison dipole. This corresponds to a power gain of over 10, making your CB station legally equivalent in power to an illegal CB station running 50 watts input into a dipole antenna!

This addition of gain with extra parasitic elements is not infinite, however, and after five or six elements have been added, tuning of the Yagi beam becomes rather sticky -- adjustment is difficult -- and such

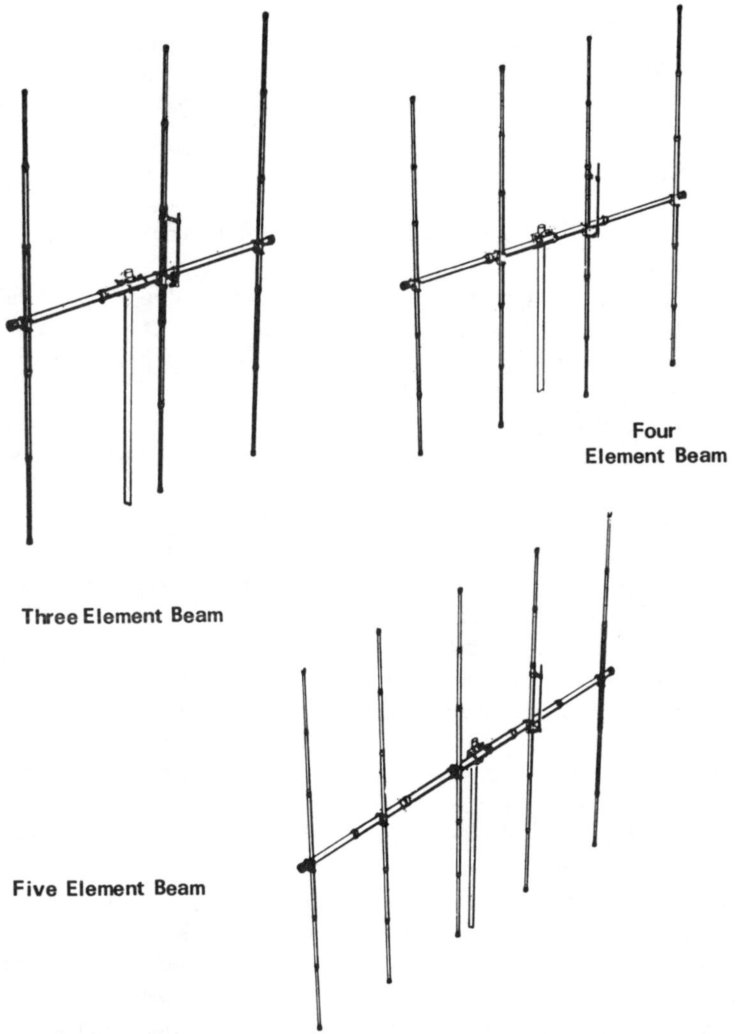

Fig. 2 THE YAGI BEAM ANTENNA is composed of a dipole driven element and a number of parasitic elements. Shown here are 3, 4 and 5 element Yagi beams, mounted in a vertical position on a supporting mast. The beam patterns are directed to the left. The driven element of each beam is characterized by the matching transformer, seen as a short section of vertical tubing attached to the dipole just above the center. Elements to the left of the dipole are directors and the element to the right of the dipole is the reflector.

Most Yagi beams are made of light weight aluminum tubing. The tubing is cut into short sections for ease of shipment and the sections telescope together and are locked in position by means of clamps or sheet metal screws.

giant antennas are not very practical, except at very high frequencies where the Yagi size is quite small and laboratory adjustments may be made easily. Antenna gain will be discussed in Chapter 7.

The Yagi may be mounted with the elements parallel to the earth or in a vertical position, or it may be cross-polarized as described in the next section.

The Cross-Polarized Yagi Beam

Two Yagi beam antennas may be mounted on the same boom with the elements at right angles to each other. By proper switching, the array may be used either for vertical or horizontal polarization, with one set of Yagi elements operating at a time. Such an antenna is often called a *crossed dipole beam,* an *X-beam,* a *criss-cross* array, or some such fanciful name. A cross-polarized Yagi is shown in Figure 3.

In most instances, separate coaxial lines are brought down from each of the two dipole elements and the operator has the choice of polarization by switching from one line to another. As only one beam is used at a time, the gain of such an array is equal to the gain of a single beam and the gain should not be counted twice (see Rule #7 of the Seven Ground Rules, Chapter 7). That is to say, if a Yagi beam has 7.5 decibels gain, two Yagi beams mounted in a crossed dipole array will still only provide 7.5 decibels gain in either horizontal or vertical polarization.

Some *cross-polarized* arrays are provided with a matching harness of special coaxial line so that both beams may be operated simultaneously. In this case, the power is divided equally between the beams, each beam receiving 50 percent of the transmitter power. In each polarization, therefore, power is down 50 percent, or 3 decibels. Since both polarizations are used simultaneously, it is possible that the 3 decibel difference is not noticed because of random reflection and refraction of the radio wave as it travels from transmitter to receiver.

Cross-polarization is handy to have in some cases. If two stations are using horizontal polarization, they can realize up to a 20 decibel improvement as compared to vertically polarized interfering signals on the channel. This can be a big advantage on a crucial communication circuit.

The All-Metal Yagi

Most Yagi beams are of all-metal construction, including the supporting structure, called a *boom.* The parasitic elements may be mount-

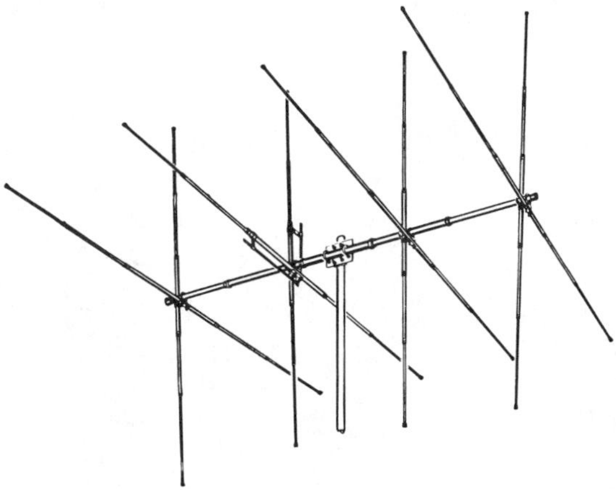

Fig. 3 THE "DOUBLE CROSS" YAGI beam antenna. Two Yagi beams mounted at right angles to each other on the same boom can be used for either vertical or horizontal polarization. Fed with two transmission lines (one to each dipole), the beam may be switched between vertical or horizontal modes by the user, by means of a coaxial switch located at the operating position. Only one beam is used at a time, and the overall gain of the array is only equal to the gain of a single beam. Switching from one polarization to the other often reduces unwanted interference and fading of the desired signal.

ed directly to the metal boom at their center point as the radio frequency current does not flow from the parasitic element into the boom. Since the dipole element is usually fed at the center, often by a two conductor transmission line, it is usually necessary to insulate the dipole from the boom by means of nonconductive material such as plastic or micarta. Some Yagis are designed so that the dipole element may be mounted directly to the boom, and some of these techniques are discussed later in the Handbook.

There's no reason why a Yagi beam could not be made out of wire instead of tubing. However, since the beam is highly directional, it is advantageous to be able to rotate it so that the beam "squirts" the signal in the desired direction. A wire beam would require some sort of supporting framework, and this would be difficult to rotate. Some radio amateurs use Yagi beam antennas slung between masts or trees, but the great majority of Yagis are made of self-supporting aluminum tubing, such as the one shown in Figure 4.

Fig. 4 FOUR ELEMENT ALL-ALUMINUM BEAM. Vertically polarized Yagi beam is popular CB antenna. This antenna has parasitic directors at left and reflector at right. The driven dipole element has a matching transformer rod visible above horizontal support boom. Yagi beam may be mounted with elements in vertical or horizontal plane. Majority of CBers employ vertical polarization, although horizontal polarization is gaining in popularity because of increased rejection of auto ignition noise.

Stacked Yagi Beams

Two beam antennas may be placed side by side, or one over the other, to double the power obtained from a single beam. Antennas connected in this fashion are said to be *stacked or phased*. A stacked three element array is shown in Figure 5. Assuming the gain from a single array is 8 decibels over a dipole, the gain from two stacked arrays is twice the power (8 + 3 = 11 decibels) over a dipole, or 13.1 decibels over an isotropic radiator. (Remember that a power gain of two equals 3 decibels). The additional gain is achieved because the beam pattern is narrower, and less energy is wasted in unwanted directions.

Yagi beams may be stacked in a vertical or a horizontal position, and the minimum stacking distance is about half-wavelength, or about

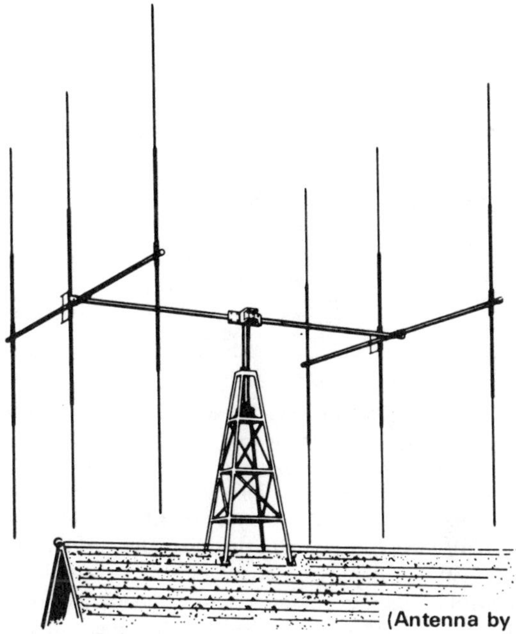

Fig. 5 A TWIN YAGI BEAM. Two beams placed side by side ("stacked") provide twice the power gain of one beam. Stacking distance is about one-half wavelength (18 feet for the CB channels). The boost signal is achieved by "squeezing" the beam pattern. For correct operation, the two antennas are fed in parallel by a phasing harness. This is simply a length of coaxial line running from each beam to the station equipment. The lines must be of equal length. At the station, the inner conductors are connected together as well as the shields, placing antennas in parallel connection.

18 feet at CB frequencies. Stacking distances down to as low as 9 or 10 feet are sometimes used, at the expense of stacking gain. Since stacking gain, at the best, is only about 3 decibels (double the power), it is important that gain not be sacrificed by reducing the stacking distance.

Part II. The Cubical Quad Beam Antenna

Of great interest to CBers and radio amateurs alike is the *Quad* beam antenna, so-called because of its square shape. Originally designed by a radio ham, W9LZX, and first used at the shortwave broad-

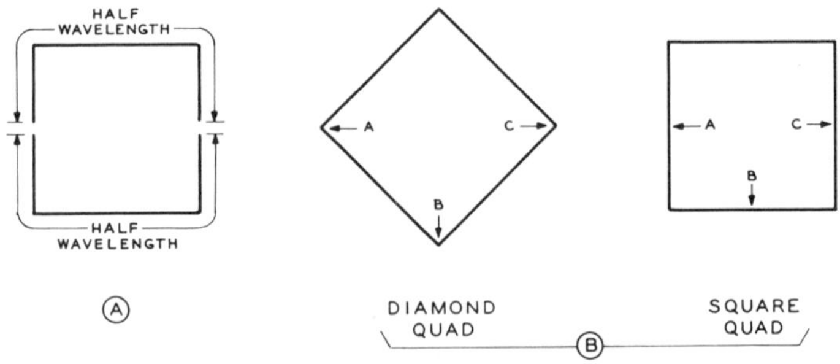

Fig. 6 BASIC QUAD ELEMENT may be thought of as two half-wavelength antennas placed one above the other and interconnected by the tips, which are folded inwards. A typical Quad element is about one quarter wavelength on a side, or one wavelength in circumference. The Quad element may be either diamond or square in shape (B) with little difference in performance between the configurations. For vertical polarization, the driven loop should be fed at either A or C. For horizontal polarization, the driven loop is fed at point B. Parasitic elements, having no feedpoint, "do not know" the polarization of the driven element and work equally well with either polarization. See Figure 9 for feed systems.

cast transmitter of HCJB in Ecuador, the fame of the Quad has spread until today it is one of the most popular amateur antennas. It is, moreover, gaining popularity among CB operators.

The Quad antenna is inexpensive to build and provides excellent gain for its small size. It is easy to adjust and almost foolproof in operation. As shown in Figure 6A, the Quad may be thought of as two Yagi antennas, placed one above the other and interconnected by the tips of the elements, which are folded in towards each other.

The Quad is a tricky antenna to build as it is a three dimensional object, having length, width and height, and is relatively hard to handle on the ground. It has an affinity for getting tangled up in guy wires during the erection process and can easily shake itself to pieces in a strong wind unless expertly constructed. Only an experienced (or ambitious!) antenna builder should attempt a Quad antenna project, as the mechanical problems are formidable. Luckily, various manufacturers make Quad kits which supply some of the more complicated mounting hardware. For those of a mechanical nature, the Quad is a rewarding construction project and is the "antenna to beat" in the worlds of CB and amateur radio.

The simplest Quad is the single loop (Demi-Quad), which provides a power gain of about 2 decibels over a dipole. It is shorter in "wing spread" than the dipole (being about 9 feet on a side for CB frequencies) and has a figure-8 pattern similar to that of a dipole (Figure 6B). Depending upon whim, the Quad may be made either rectangular or diamond shaped, as shown in the illustration. For these designs, the two wire transmission line is attached to the loop, which is broken at one point.

The most popular Quad design is the two element Quad, making use of a second loop as a reflector element. The two element Quad provides a power gain of about 7 decibels--nearly as much as a full-size three element Yagi beam. Since the "wingspread" of the Quad is about half of that of the Yagi, the two element Quad is popular for those CBers with restricted antenna space who wish to have an effective beam antenna (Figure 7).

The three element Quad, composed of the dipole loop, a reflector and a director, while quite practical, is not too popular since the center loop interferes with the mounting assembly and gets in the way of the supporting tower or mast.

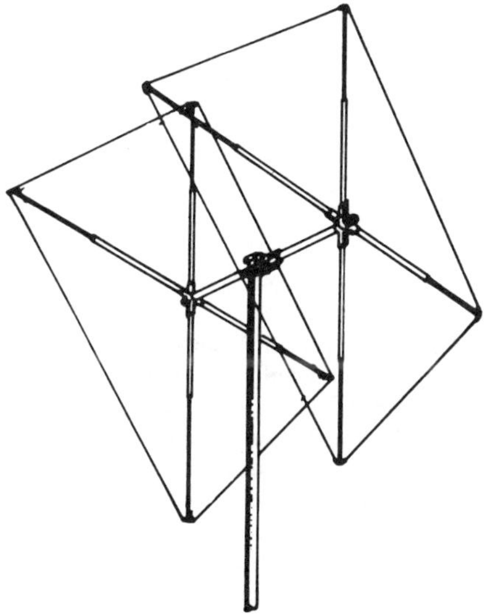

Fig. 7 TWO ELEMENT QUAD is a popular beam for both CB and ham operation. Elements are made of wire strung on insulated element tips. This particular design has aluminum crossarms with two foot tips made of phenolic material which support element wires clear of the metallic structure. Quad beam may be either square or diamond shaped with equal performance.

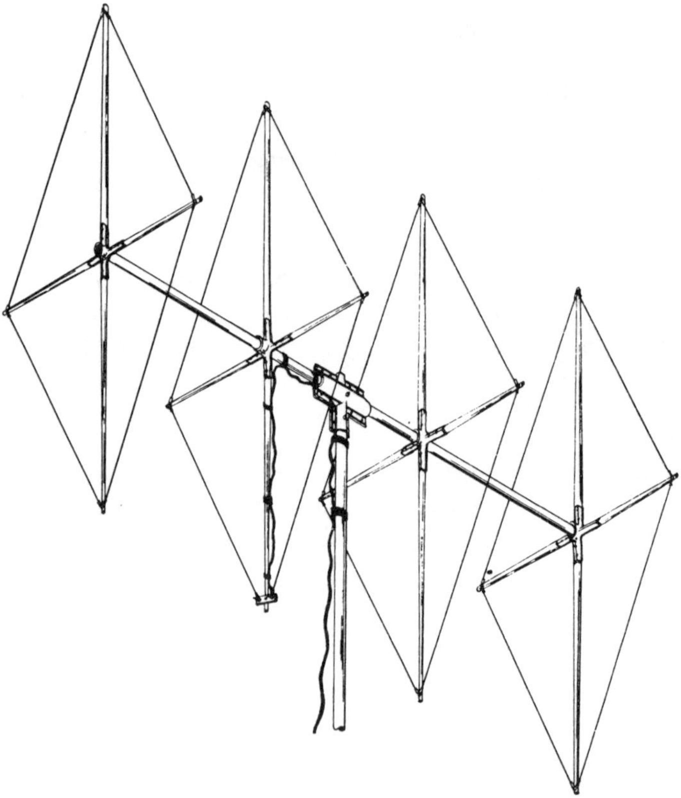

Fig. 8 THE MONSTER QUAD! Four element Quad antenna provides highest power gain per unit of size than any equivalent CB antenna. Either diamond or square configuration provides over 14 decibels gain as compared to an isotropic radiator (12 decibels gain over a dipole antenna). This corresponds to a power gain of 15, making your 5 watt CB transmitter equivalent to one having a power input of 75 watts! The four element Quad may be brought from various manufacturers in kit form, or built from complete information given in this Handbook. The Quad shown above is constructed on a 16 foot aluminum boom with the element arms made of aluminum tubing and phenolic insulating tips. Other designs use Fiberglas arms. The beam pattern of the four element Monster Quad is quite narrow and the antenna must be accurately aimed for best results. Because of weight and wind resistance of Monster Quad, a heavy duty rotator must be used.

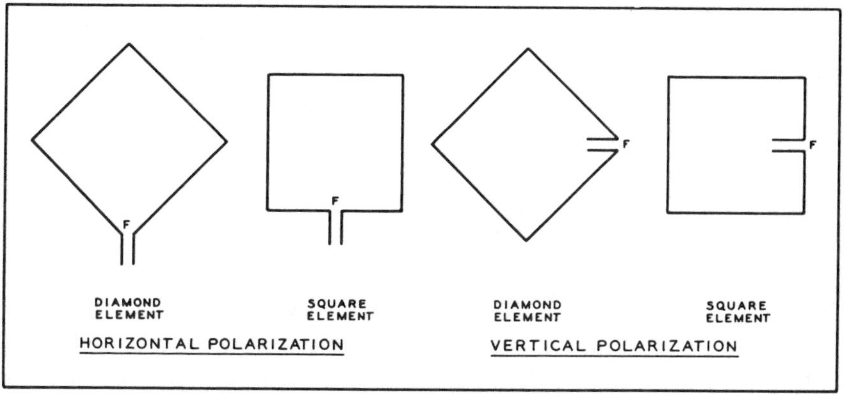

Fig. 9 HORIZONTAL OR VERTICAL POLARIZATION of Quad antenna is determined by placement of feedpoint (F) in the loop. Parasitic elements remain the same regardless of polarization of driven element. Most CB Quads are vertically polarized so as to be compatible with vertical polarization of other CBers. Some operators prefer horizontal polarization to reduce interference from vertically polarized stations.

The antenna causing the most excitement on the short waves is the four element Quad which is composed of the dipole loop, a reflector and two directors (Figure 8). This "Monster Quad" provides over 12 decibels gain compared to a dipole, and is about the largest practical beam antenna for CB and amateur use. Construction of a "Monster Quad" is covered later in this Handbook. For those CBers who have the time and patience to build this excellent antenna -- and who have the space to erect it -- it will provide superlative results for transmitting and receiving and make the 5 watt CB station sound like it is using 75 watts!

Feeding The Quad

The Quad antenna may be either vertically or horizontally polarized by proper placement of the feedline on the driven loop as shown in Figure 9. For most CB work, vertical polarization is used. In this case, the Quad is fed at the mid-point of one of the sides. It does not make a difference which side is chosen. For horizontal polarization, the Quad is fed at the mid-point of the bottom of the loop.

Coaxial transmission line (RG-58/U) may be used to feed the Quad antenna through a simple *balancing transformer* (balun) mounted at the Quad. Construction of an inexpensive balancing transformer will be discussed in a later chapter of this Handbook.

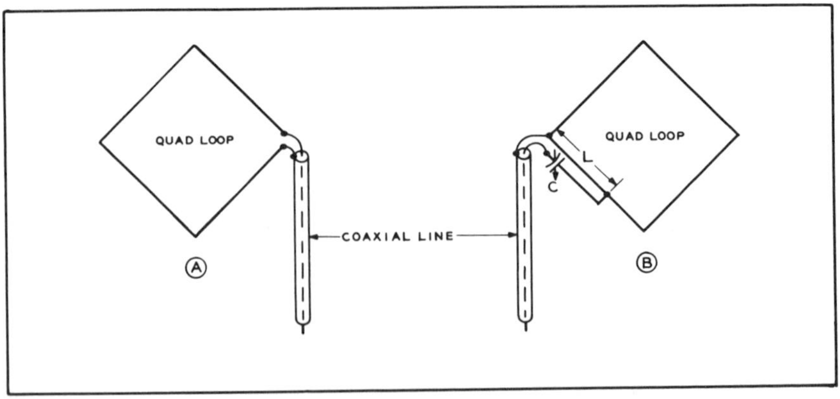

Fig. 10 DIRECT FEED for Quad loop is shown at (A). A balun may be placed between coaxial line and loop, if required. Closed-loop feed system with Gamma match is shown at (B). Length of matching rod (L) and value of capacitor (C) determine match. See Fig. 7, page 204 for details.

Quad Feed Systems

Regardless of polarization, most CB Quad beams are either fed directly (with or without a balun) or with a Gamma match system (Fig. 10). The Gamma match permits the operator to compensate for minor variations in the electrical characteristics of the Quad caused by the construction techniques used and also allows the resonant frequency of the antenna to be moved above and below the natural resonant frequency. More on the Gamma match on pp. 203-204.

Fig. 11 CLEVER 3 ELEMENT QUAD uses diamond shaped elements. Driven element is set off-center on boom so that lower portion of wire loop does not hit the supporting mast as the Quad is rotated. Fibreglas elements are used in this efficient antenna.

Quad Construction

Quad construction is quite different from that of the Yagi antenna. The most popular arrangement is to make a cross-type framework that supports the wire loops. The framework is often made of bamboo poles, or of the more expensive, durable Fiberglas poles. Some manufacturers make the Quad support structure of aluminum tubing. When this is done, the structure should be broken by insulators at various points to prevent it from acting as an antenna itself and distorting the gain pattern of the Quad. The supporting boom of the Quad is usually made of aluminum tubing, and special support fixtures are available which permit the insulating arms to be affixed to the metal boom, as shown in the illustration.

Beam Pattern of the Quad

Generally speaking, the beam pattern of the Quad resembles that of the Yagi, although--element for element--the pattern of the Quad is "tighter" (narrower) than that of the Yagi. A "Monster Quad" must be aimed quite accurately, as the beam pattern is very sharp.

WHO NEEDS A BEAM ANYWAY? This CBer attracts plenty of attention with her one watt walkie-talkie. When she talks to a base station having a beam, her radio range is extended. Aren't you glad you have a beam?

Chapter 7

The TRUTH About Antenna Gain

(P.T. Barnum is Alive and Well --

and Writing Advertisements for CB Antennas)

To overawe the gullible and impress the prospective buyer, a veritable barrage of non-facts has been invented and used by a small number of antenna manufacturers who hope to capture the lion's share of the CB antenna market. OK-sounding but meaningless terms such as "directive gain", "talk power", "input gain", "effective DX gain", "radiated gain", "gain proven by U.S. Patent", and other nonsense terms are used to extoll the virtues of antennas manufactured by the clever to be sold to the unsuspecting CBer.

The great majority of antenna manufacturers are honest but the barrage of bull-dust from a few quacks baffles the brains of those trying to select an efficient CB antenna and it is a continuing source of amazement and frustration to serious, competent manufacturers who truthfully advertise their products in the face of such humbug.

Unfortunately, many garage-shop manufacturers, realizing that the majority of CBers are nontechnical types, make a great and continuing effort to "snow" them by the use of unidentified terms and imposing claims in advertising their antenna products. This state of affairs is verified by the editors of one CB magazine who state that their greatest amount of mail comes from readers asking questions about various "wonder antenna" advertisements that boast of exotic gain.

Power Gain: The Name of the Game

The most sought-after quality in an antenna of any type is power build-up or *power gain*, achieved by focusing the radio waves in one direction. Power gain, as noted previously, is formally referred to in decibels (or db), with zero decibels being the power level of a reference or comparison antenna, against which the "miracle" antenna can be compared.

Watch out! You'll read many references to power gain in antenna advertisements, along with big, impressive numbers. Colorful statements such as "this antenna gives you 30 watts talk power", sound exciting, and thrilling words like this are common among the humbug artists.

Let's face it. The antenna manufacturer is competing for your hard-earned dollars just as is the automobile dealer who boasts of horsepower or the stereo salesman who extolls the audio quality and "music power" (often unneeded) of his amplifier. Like other members of this hard-sell family, a few antenna manufacturers sometimes let their enthusiasm for the product run away from the facts. As a result, antenna specifications are often written in the advertising department, not in the engineering laboratory!

So--a word to the dollar-wise CBer: *forget* imposing and meaningless terms extolling antenna gain that are designed to separate you from your hard-earned money. There's a better way of judging CB antennas, a "yardstick" that has proven itself over the years to be fair and accurate. You can apply this "yardstick" using the same ground rules that experienced radio engineers and communicators apply to their antennas, *for all antennas operate by the same rules of nature.*

The following suggestions (compiled by the authors and knowledgeable engineers in the antenna business who have no axe to grind) provide you with a "yardstick" of measurement and a unique *Truth Table* which you can apply to *any* CB antenna, regardless of type or manufacturer. By referring to the "yardstick" guidelines and the Truth Table, you will achieve a good knowledge of antenna power gain performance, regardless of advertised performance figures given for the antenna in question.

What's The Truth About Antenna Gain?

Antenna power gain in the desired direction is obtained at the expense of signal radiation in other directions. Thus, a radio wave may be focused in a direction so that the signal will *sound* as though it comes from a much more powerful source. The directional beam antenna may be rotated through 360 degrees to "squirt" the signal in the wanted direction, just as light may be focused and pointed about with a flashlight. Such a beam antenna can boost the transmitted signal and improve the received signal over the performance of a more simple antenna. Some beam antennas provide more gain than others, but the basic theory of operation of beam antennas is well known and is covered in the literature. For technically minded CBers, a list of good technical books covering antenna theory and design is included at the end of this chapter.

To begin with, a good rule-of-thumb to remember when investigating any antenna is this: *the gain of an antenna is proportional to the size of the antenna when compared with the size of the radio wave.* Simply: big antennas tend to have big gain, and little antennas tend to have little gain. Miniature antennas (coil loaded antennas, for example) may actually exhibit signal *loss*, instead of signal gain!

In addition, just because a particular type of antenna has a maximum *theoretical* gain figure, it is often risky to assume that the antenna made by a particular manufacturer exhibits all of this possible gain. Manufacturing imperfections and design goofs can severely limit power gain. If the manufacturer doesn't know what he is doing (and some of them don't), the beam antenna can actually exhibit a negative power gain, or signal loss!

Designing and building a good, efficient, mass-produced beam antenna to sell at a competitive price is not easy. The larger, reliable manufacturers have extensive antenna test ranges and perform exacting tests and measurements on their antennas. Even then, they are under constant pressure from competition to "sell" the performance of their products strongly because of the fear of losing sales to companies whose advertising scruples are more elastic and who are not afraid to exaggerate, knowing that the average CB operator is not an antenna engineer. This makes it tough for everybody; the reputable manufacturer is at a disadvantage because his true gain figures seem poor in comparison with the inflated figures of his competitor, and the prospective buyer is completely confused as to the actual performance of both antennas.

The problem of inflated antenna performance claims is well known and various organizations (such as the *Electronic Industries Association)* are making attempts to standardize antenna information. Some radio amateur magazines refuse to accept any antenna advertising which contains gain figures until such standards are established, since so much abuse exists in this area. CB magazines may well do this at some future date. Meanwhile, *caveat emptor - "let the buyer beware".*

The "Yardstick" of Performance

Antenna power gain is expressed in decibel units (or db) and is referenced against a standard comparison antenna. The comparison antenna used in industry is the *isotropic radiator.* In order to understand the concept of this unique antenna, you should imagine the antenna to be inclosed in a hollow sphere, as shown in Figure 1. If the radiation from the antenna is distributed uniformly over the interior surface of the sphere, the antenna is said to be isotropic. Any antenna which causes

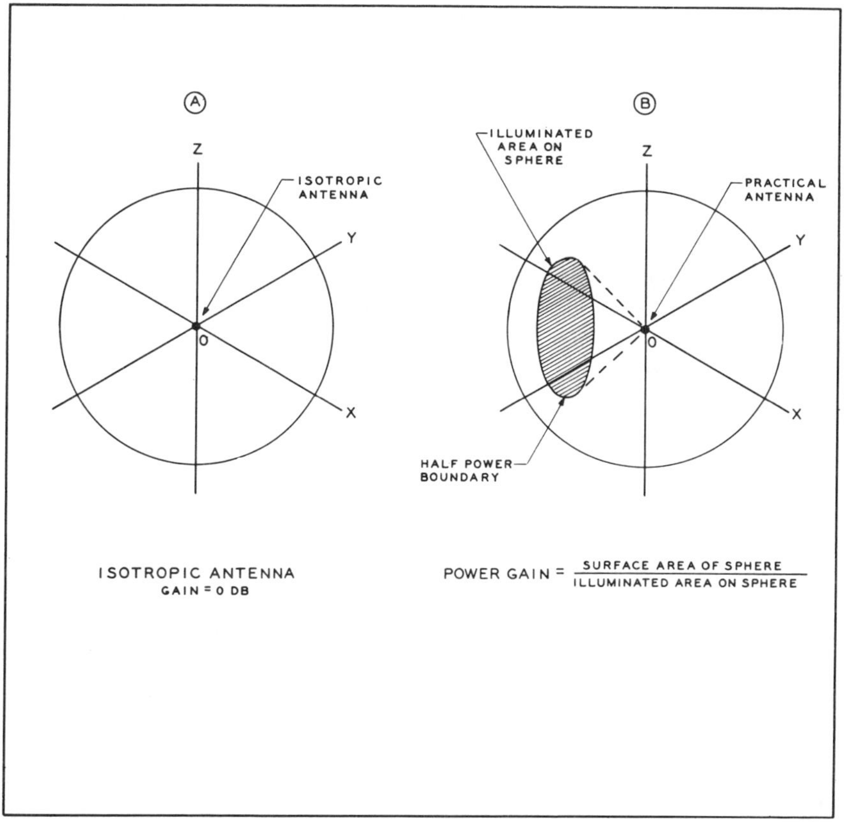

Fig. 1 ISOTROPIC ANTENNA is point source which uniformly illuminates interior surface of a sphere. An antenna which concentrates radiation into a certain area of the inside surface of the sphere, and which produces greater intensity than that produced by an isotropic radiator fed with equal power, is said to exhibit gain relative to an isotropic radiator. Power gain is inversely proportional to fraction of total interior surface area which receives the concentrated radiation. For purpose of comparison, imagine the isotropic antenna to be a candle placed at the center of a large, inflated balloon. Practical antennas illuminate the sphere unequally and all of them exhibit power gain over an isotropic radiator. The area of illumination is not sharply defined as shown by the shaded region, but gradually falls away from the center of the area. By definition, the boundary of the area is determined by the line at which the radiation intensity has fallen by half (3db). The boundary is called the "half power" point. The illuminated area can be plotted graphically and such a plot is called a "polar diagram". A typical polar diagram is shown in Figure 1, Chapter 6.

the radiation to be concentrated into a particular area of the inside surface of the sphere, producing a greater intensity than that produced by an isotropic radiator fed with equal power, is said to exhibit *power gain with respect to an isotropic radiator*. The gain, moreover, is inversely proportional to the fraction of the total interior surface area which receives the concentrated radiation.

Since gain is expressed in relative terms, it is very useful to express the gain of any two antennas in terms of a standard antenna. The performance of any two antennas then may be directly assessed by comparing their relative gains over the standard antenna. While the isotropic radiator is considered to be the standard comparison antenna, it is a non-practical standard, which cannot be constructed. Its value lies in the fact that it does not require a definition of the direction of radiation for a comparison to be made, as the radiation is nondirectional.

Three standard antennas are commonly used as a "Yardstick of Comparison" and each provides a slightly different gain figure. By not stating the antenna used for comparison, some manufacturers can play games and make their advertised gain figures look better than they are.

The isotropic radiator, as mentioned before, is the basic standard of comparison in antenna engineering work. It has a power gain of one (unity), or *zero decibels*. An approximation of an isotropic radiator could be a very short section of a full-size dipole antenna (perhaps only 1/100th wavelength long) with constant current along its length and an efficiency of 100%. The isotropic radiator, because of its small size, has a "negative" power gain of -2.1 decibels when compared against a standard half-wave dipole antenna. Or, to put it another way, the half-wave dipole has a power gain of 2.1 decibels over an isotropic radiator.

The half-wave dipole is the 2nd comparison antenna. It has a power gain of 2.1 decibels over the isotropic radiator, since it concentrates energy broadside to the dipole instead of "spraying" it in all directions.

A third comparison antenna is the ground plane antenna. It has a power gain of about 0.3 decibel over the isotropic radiator. The gain figure of the ground plane varies a bit, depending upon placement of the radials, as is discussed later. However, this gain figure will be accepted as "par" for this "Yardstick of Comparison".

It is apparent that the published gain of an antenna can vary as much as 2.1 decibels by merely changing the comparison antenna from a dipole to an isotropic antenna.

Which of these three antennas is used as the comparison antenna when the manufacturer states the gain figures for his antennas? You'll never know, unless he tells you! His gain figures, then, are meaningless unless the reference antenna is noted, since each reference antenna provides a different starting point on the gain scale. However, by deduction, and the use of the Truth Table, you can determine what the power gain figure in decibels should be for any antenna, *regardless of what the manufacturer tells you!*

The Seven Ground Rules

The following seven ground rules provide you with an accurate "yardstick" of performance for any antenna. Here they are, using the theoretical isotropic radiator as the reference point of performance:

Rule #1 - The fundamental comparison antenna is the isotropic radiator and its gain is zero decibels.

Rule #2 - The gain of a half-wave dipole antenna is 2.1 decibels over the isotropic reference antenna.

Rule #3 - The gain of a single Quad loop driven element is 4.1 decibels over the isotropic reference antenna.

Rule #4 - The gain of an antenna which has a *single* parasitic reflector or director element is 5 decibels over the gain of the antenna driven element itself.

Rule #5 - Additional parasitic *directors* provide additional gain in a decreasing manner. One additional director element (over the first director) provides 2 decibels gain. Each additional director thereafter adds about 1 decibel additional gain.

After 4 or 5 directors are added, the improvement in gain is negligible. (Only one reflector element is effective in a simple beam antenna so adding additional reflectors produces no additional gain).

Rule #6 - When both parasitic reflector and director(s) are used, the *reflector* gain figure is reduced from 5 decibels to 3 decibels.

Rule #7 - Antenna gain measured in the vertical plane cannot be counted again in the horizontal plane, and vice versa.

OK, gentlemen, these are the ground rules. Our basic comparison antenna is the imaginary theoretical isotropic radiator. Let's make these rules work for you!

Example: What is the power gain of a 3 element Yagi beam?

Answer: The 3 element Yagi beam is made up of a dipole (the driven element which is "excited" by the transmitter), a reflector element,

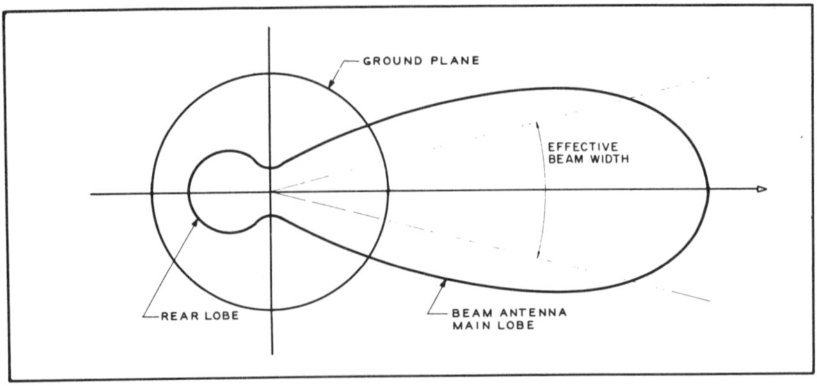

Fig. 3 POLAR PLOT shows antenna radiation pattern as seen from above. Shown here are typical polar plots of nondirectional ground plane and directional beam antenna. The directive pattern of the beam is called the "main lobe". Beam width of main lobe is expressed in degrees and is measured at points on pattern at which power is reduced by one-half (-3 decibels). Thus, for example, if the effective beam width is 28 degrees, the main lobe is said to be "28 degrees wide at the -3 db points". All beams exhibit some radiation to rear of array,

and a director element. We are making comparison against an isotropic radiator (Rule #1) so that is our reference, or starting point. Rules #2, 4, and 6 apply.

The power gain is:

Dipole gain	=	2.1 db	(Rule #1)
Reflector gain	=	3.0 db	(Rule #4 and #6)
Director gain	=	5.0 db	(Rule #4)
Antenna gain	=	10.1 db	compared to an isotropic radiator

Example: What is the power gain of a 4 element Quad beam?

Answer: The 4 element Quad beam is made up of a Quad loop driven element, a reflector element, and two director elements. We are making comparison against an isotropic radiator (Rule #1) so that is our reference, or starting point. Rules #3, 4, 5, and 6 apply .

The power gain is:

Quad loop element gain	=	4.1 db	(Rule #3)
Reflector gain	=	3.0 db	(Rules #4 and #6)
First Director gain	=	5.0 db	(Rule #4)
Second Director gain	=	2.0 db	(Rule #5)
Antenna gain	=	14.1 db	compared to an isotropic radiator

Example: What is the power gain of a 2 element Yagi beam using a director?

Answer: The 2 element Yagi beam is made up of a driven element and a single director. We are making comparison against an isotropic radiator (Rule #1), so that is our reference, or starting point. Rules #2 and 4 apply.

The power gain is:

Dipole gain =	2.1 db	(Rules #1 and #2)
Director gain =	5.0 db	(Rule #4)
Antenna gain =	7.1 db	compared to an isotropic radiator

Got the idea? Antenna gain is determined simply by adding up the individual gain figures according to the Ground Rules.

The Dipole Antenna as Reference

Some antenna manufacturers publish gain figures for their beams that are referenced to a dipole antenna instead of an isotropic radiator. What happens to the Ground Rules in this case? Simple. Omit Rule #1, thus subtracting 2.1 decibels from the total gain figure.

Thus, in the first example, the power gain of a 3 element Yagi beam is estimated to be 10.1 db over an isotropic radiator. The gain figure of this antenna, compared to a dipole, is 2.1 db less, or 8.0 db. This comes about since using the dipole for comparison means that the starting point for computation is 2.1 decibels higher than if the isotropic radiator was used for the starting point.

Likewise, in the second example, the power gain of a 4 element Quad beam over a reference dipole is 2.1 db less than when compared to an isotropic radiator. Specifically: 14.1 - 2.1 = 12 decibels gain over a comparison dipole.

Finally, in the third example, the power gain of 2 element Yagi beam using a director over a reference dipole is 5 decibels.

Comparison Between Two Different Antennas

The Ground Rules also may be used to compare one antenna against another. This is done in the same fashion, using the Ground Rules and adding gain figures. Strictly speaking, vertical and horizontal antennas cannot be compared against one another as far as actual gain goes, but such a comparison is meaningful in terms of results, and provides the operator with a real-life comparison between vertical and horizontal polarization.

Example: How much additional gain does a 4 element Quad show over a 3 element Yagi beam?

Answer: The gain of the 4 element Quad is determined to be 14.1 db over an isotropic radiator. The gain of the 3 element Yagi beam is determined to be 10.1 db over isotropic. The 4 element Quad has an advantage of 14.1 - 10.1 = 4 decibels over the Yagi.

Rule #7 -- What Does It Signify?

Rule #7, "Antenna gain measured in the vertical plane cannot be counted again in the horizontal plane, and vice versa", has not been mentioned so far. This rule applies to "cross-polarized" antennas or any antenna type that may be operated with either horizontal or vertical polarization. For example, if a certain antenna has 6 decibels power gain in the horizontal position, and also 6 decibels power gain in the vertical position, it cannot be said that the total antenna power gain is 12 decibels. Some antennas, moreover, have complex feed systems that permit radiation in both horizontal and vertical planes. In this case, the power is divided equally between the planes, each plane receiving one half the transmitter power. Simply, the rule means that antenna gain cannot be counted twice.

Four Ground Rules For Vertical Antennas

A second set of four Ground Rules applies to vertical antennas such as the ground plane, the half-wavelength vertical, and the 5/8-wavelength vertical, as these antennas are not beams in the true sense and have no parasitic elements. The rules are:

Rule #1 - The fundamental comparison antenna is the isotropic radiator and its gain is zero decibels.
Rule #2 - The gain of a ground plane antenna is 0.3 decibel over the isotropic reference antenna.
Rule #3 - The gain of a vertical half-wavelength (dipole) antenna is 2.1 decibels over the isotropic reference antenna.
Rule #4 - The gain of a 5/8-wavelength vertical antenna is 3.3 decibels over the isotropic reference antenna.

And how do you tell the difference between these vertical antennas, as they all resemble a vertical stick? Simple. Measure or estimate the overall height. In the CB bands, the ground plane is about 9 feet high,

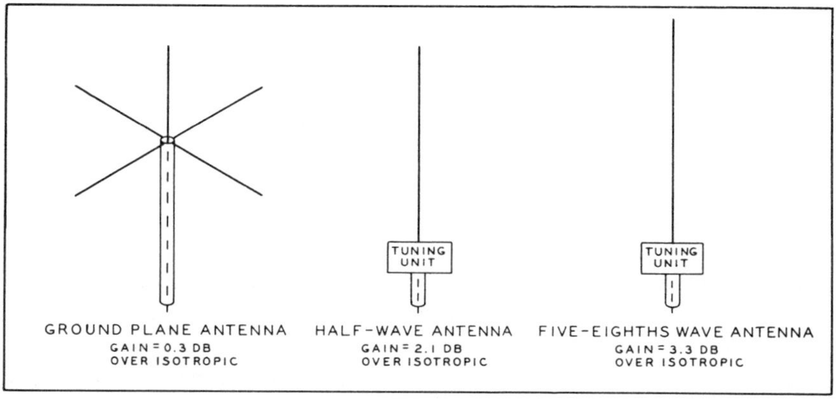

Fig. 4 GROUND PLANE is often used as standard of comparison for vertical antennas. Half wave vertical provides power gain of 1.8 decibels over ground plane and five-eights wave vertical provides power gain of 3 decibels over ground plane. The five-eighths wavelength antenna is recommended for general base station use.

the half-wavelength antenna is about 18 feet high and the 5/8-wavelength antenna is about 20 feet high.

Working from the above Ground Rules, it can be seen that the power gain of a vertical half-wavelength antenna is 1.8 decibels over a ground plane, and the power gain of a 5/8-wavelength vertical antenna is 3.0 decibels over a ground plane.

Radial Placement

Some ground plane antennas have "drooping" radials. That is, the radials are not in the horizontal plane, but drop away at about a 45 degree angle. This is often done to make a more correct impedance match between antenna and transmission line, or to make the radials serve as supports for the antenna structure. When the radials are allowed to droop, the gain of the ground plane is raised by about 0.5 decibel over the normal configuration and the antenna begins to resemble a sleeve-type half-wave dipole. It is doubtful if this slight increase in power gain could be noticed by a casual observer listening to a signal transmitted first from a conventional ground plane and then from a "drooping" ground plane.

Generally speaking, placement of the radials or their number is not particularly critical. Most ground plane antennas have four radials but experience has shown that three radials work just as well.

A Word of Caution

Measuring antenna gain is a complex science and gain measurements accurate to one decibel or better can only be made on expensive and extensive antenna measuring ranges. Only a few manufacturers possess such facilities. Gain figures, therefore, should be taken as a measure of theoretical performance and with a large lump of salt, and not accepted as gospel. Unfortunately, antenna power gain is a fetish, it seems, and is often used as powerful advertising medicine.

To help you interpret antenna gain, a unique Truth Table has been devised which places antennas in their proper relationship as far as power gain is concerned. It is covered in detail in the next chapter.

Recommended Reading

Here is a list of books that deal exclusively with antenna systems. They are on a technical level higher than this Handbook and are recommended to all CBers with a serious interest in antennas.

Orr and Cowan, "Beam Antenna Handbook", Radio Publications, Inc., Box 149, Wilton, CT 06897. (All about Yagi beams in simple terms).

Orr and Cowan, "All About Cubical Quad Antennas", Radio Publications, Inc., Box 149, Wilton, CT 06897. (Full information on Quads).

---"ARRL Antenna Handbook", American Radio Relay League, Inc., Newington, CT 06111. (Practical data on amateur antennas).

"Tell me just once more about isotropic antennas......."

Chapter 8

The TRUTH TABLE

(Things Your Mother Never Taught You About CB Antennas)

All radio antennas follow natural rules that are well known to electronic engineers. Antenna power gain may be predicted mathematically before an antenna is ever built and the theory may be proven with a model antenna tested on an antenna range. Very seldom does a "miracle" antenna come along which obsoletes the great sum of knowledge built up over the years since Hertz first used his "sparking ball" antenna.

The most important antenna information covering power gain may be summarized in the "yardstick" and Ground Rules discussed in the previous chapter. These are classified into a unique Truth Table that provides at a glance the theoretical maximum power gain figure of the most popular antenna arrays. Using the table, you can quickly determine the gain performance of an antenna as compared to an isotropic radiator, or to a dipole, or to another antenna on the Truth Table. This helps you to sift out impossible advertising claims, inflated gain figures, and other delightful practices designed to confuse you.

The Truth Table

The Truth Table summarizes antenna power gain for the ten most popular types of CB antennas, and compares their gain in decibels. Let's see how the Truth Table works.

The first vertical column in the Truth Table lists the ten most popular antenna types, and starts with the imaginary isotropic radiator and progresses through various antennas up to the 4 element "monster" Quad, the King of CB beam antennas.

The second column *(Decibel Gain Over Isotropic Radiator)* compares the power gain of any antenna in column one to an isotropic radiator. Thus, the 3 element Yagi beam has a power gain of 10.1 decibels over the isotropic radiator, as noted on the Table.

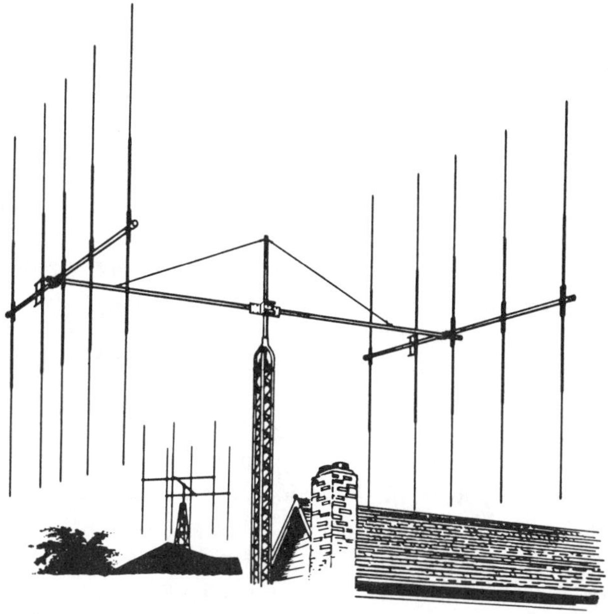

Fig. 2 ABSOLUTE GAIN OF BEAM ANTENNA is doubled by combining two identical beams in "stacked" array. Beams may be stacked in either vertical or horizontal position. Single 5 element beam has a power gain of about 11 dB over dipole and two beams, properly stacked, provide twice the power gain of one, or 11 + 3 = 14 dB. The beams are fed in parallel by a phasing harness which is composed of two equal lengths of coaxial line running from the beams to the station. At the station, the inner conductors are connected together, as well as the shields, placing antennas in parallel.

bels into S-meter readings on your CB receiver meter. Yes, it is well known that many S-meters are calibrated in decibels, but no two manufacturers agree on what the S-meter should indicate and an S-6 signal on one receiver may be S-9 on the meter of another receiver. Yet the signal may sound the same on both receivers! In addition, the S-meter scale--in most cases--is arbitrary, with *no* correlation between actual decibels of power gain and meter reading, despite the markings on the meter. Expressions of signal strength without reference to conditions of measurement and decibel reference are--as expressed by Poo-Bah in the operetta "The Mikado"--"merely corroborative detail, intended to give artistic verisimilitude to an otherwise bald and unconvincing narrative".

In the past, some manufacturers of better CB equipment have tried to make one S-unit equivalent to 4 decibels power gain, but this definition is the exception, rather than the rule.

A word of warning, then--please don't try to interpret the Truth Table in terms of the "funny" S-meter of your CB transceiver!

The "Gain Ladder"

The Truth Table of Figure 1 may be expressed in terms of a "gain ladder" on which the various popular CB antennas are placed in relation to a zero gain position. Such a ladder is shown in Figure 3. The left-hand vertical rail of the ladder indicates antenna gain compared against an isotropic radiator and the right-hand rail indicates gain compared against a half-wave dipole reference antenna. The antenna at the top of the ladder has the highest power gain, with gain decreasing the farther down the ladder a particular antenna is positioned. Foremost in gain, and the occupant of the ladder top is the four element Quad antenna, the "King of CB Antennas". It outranks the popular three element Quad and four element Yagi beams by 2 decibels and shows a power gain of 14 decibels over a ground plane. Those CBers used to ground plane performance will be astounded at the dramatic improvement in results, both transmitting and receiving, when a Monster Quad beam antenna is used. Even so, the smaller beams give a good account of themselves as far as power gain is concerned, and the gain ladder shows in dramatic fashion why even the smallest beam antenna outperforms the simple ground plane antenna.

Before you buy or build a CB antenna, study the gain ladder and the Truth Table. Compare antenna cost against gain. Compare published gain figures for the antenna of your choice against the gain figures given in these two tabulations. It may be fun to be fooled, but being *smart* can save you time and money.

The Truth Table Pays Off!

The Truth Table is a handy device to evaluate the most popular CB antennas. Now, if a description of an antenna (or a photograph or drawing) is given to you, you can often determine the antenna type from inspection of the photograph by counting the antenna elements, noting to which element the feedline is connected, etc. Then, after the type of antenna is determined, look up the probable power gain figure for this antenna in the Truth Table. Generally speaking, no one antenna manufacturer has a monopoly on excellence, antenna power gain, or is the ultimate source of antenna wisdom or design. Power

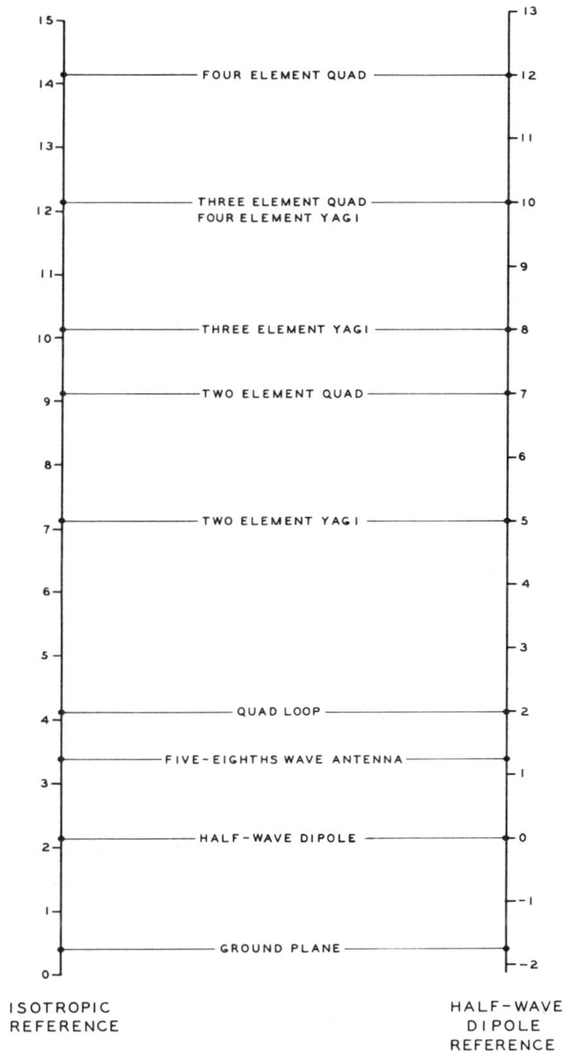

Fig. 3 GAIN LADDER shows relative power gain of popular CB antennas expressed against dipole and isotropic references. Left-hand vertical rail indicates antenna gain compared against theoretical isotropic radiator and the right-hand rail indicates gain compared against a half-wave dipole antenna. All antennas shown give good power gain over popular ground plane. The "Monster" 4 element Quad provides highest power gain of all antennas shown: 12 decibels over ground plane!

gain aside, the quality and ruggedness of antenna hardware and components are of great importance and should not be sacrificed for price or allegedly superior antenna performance. A poorly designed antenna will quickly fall apart in bad weather when hit by wind, ice, snow, etc. Poorly made or flimsy elements corrode and antenna performance suffers.

Once you have decided on a certain type of antenna, shop around. Look at antennas in use in your town and observe how they stand up under bad weather. Talk to other CB operators and to radio amateurs. Above all, speak to your distributor or dealer of CB equipment. Visit his store and examine the antennas he has in stock. Ask for his judgement. Check antenna construction and hardware. Write the various antenna manufacturers for their catalogs and full details. After all, it is your money you are spending!

On-The-Air Antenna Checks

Some experimentally minded CBers attempt on-the-air antenna checks, switching back and forth between two antennas to determine if one performs better than the other. Such tests, if properly conducted, can be valid only for the particular circumstances under which they are run. It should not be generally assumed from such a test that if antenna A works better than antenna B at one station, it will outperform antenna B at a different station location under different environmental conditions.

If you wish to run a controlled on-the-air check between two antennas, it can provide you with useful information for your individual location. The two test antennas should be in the clear and located at about the same height above ground. They should be adjusted so that the SWR on each transmission line is reasonably low. The antennas should be connected to your CB equipment through a coaxial switch, so that the two feedlines can be transferred back and forth rapidly.

A nearby friend can serve as a test observer. He should be a mile or so away from you and his antenna should have the same polarization as yours. He will note the difference in signal strength between the two antennas as you switch back and forth, making a short transmission on each antenna. The tests should be repeated several times, and at different hours during the day (preferably when the band is dead!) During all transmissions, an SWR meter should be used in the feedline of your transmitter to monitor power output and SWR on both antennas. An effort should be made to insure that nothing changes at the transmitting and receiving locations other than switching the antennas, otherwise the results of the experiment may be confusing and inaccurate.

The difference in S-meter reading between the two antennas is averaged over a series of tests, and an average difference in signal strength computed. Of course, if the S-meter calibration is unknown, the resulting difference in signal strength is not absolute. Even so, an indication of signal strength difference between the two antennas is helpful information to have at hand.

Calibrating Your S-Meter

The S-meter in your CB transceiver may be calibrated in terms of signal strength by a qualified technician. This is how he will do it. Your receiver is connected to a radio signal generator which has an 11 meter output calibrated in microvolts of signal strength. By sending this test signal into your receiver and noting the corresponding S-meter readings as the signal strength is varied, the technician can plot a graph of meter reading versus signal strength. The graph can then be converted into decibel units above an arbitrary signal level. The indications then can be marked on the scale of your meter. Using this accurate calibration, you ignore the old markings on the meter and read signal strength directly from the new decibel scale. As long as the receiver gain does not change (due to ageing of tubes or transsistors) the meter reading will help in determining true signal strength levels when antenna tests are being conducted.

Once the meter calibration job is done, you can run other antenna tests, such as front-to-back ratio and front-to-side ratio on your antenna, or the antenna of other nearby CB stations. You must be careful, though, as signal reflection from hills or other nearby objects can play tricks with your measurements. Experience in this measurement technique is a great teacher, and room still exists for the home experimenter to build, tune and adjust his own beam antennas. A calibrated S-meter certainly helps to do this job.

''You say you're from the FCC . . . . . .?''

Chapter 9

The Coaxial Cable

(Your Radio Lifeline)

Your CB antenna installation is made up of two basic parts which should aid and assist each other: the *radiating system* (the antenna itself) and the *transmission line* (the feedline, a sort of "radio hose"). Each part has its own role to play and when each is designed to work properly with the other--and does--you'll have a successful and efficient antenna installation.

The antenna, as you know, is made of electrically conducting material, usually metal such as copper or aluminum. The mechanical supports for the antenna are made of material which does not conduct electricity, such as plastic, ceramic or wood. Materials of this class are termed *insulators*. Indeed, *if* the antenna elements are supported at the correct spot (the center of a dipole, for instance), the need for an insulated support no longer exists, no insulator is needed, and the element may be attached directly to a metal supporting structure. This is a great advantage in large beam antennas as a more sturdy and windproof array can be built if relatively fragile insulators are not needed.

The "Radio Hose"

At your CB station, radio energy from your transmitter is coupled into the feedline and carried by the line to the antenna. The feedline, therefore, may be thought of as a sort of "radio hose" through which radio signals flow efficiently from one place to another. A "lossy" or inefficient feedline, which offers resistance to the passage of radio energy, may be compared to a stopped-up garden hose, which slows down and chokes up the water before it arrives at your pet rose bush. The high resistance feedline, in addition, weakens (attenuates) the

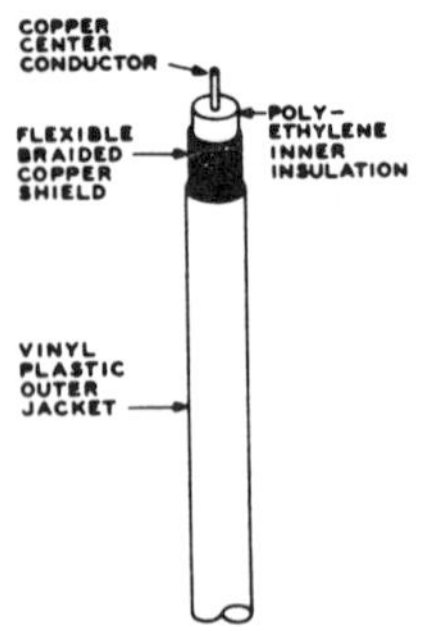

Fig. 1 COAXIAL CABLE is your "radio hose" through which radio energy flows to and from your antenna. The "coax" line consists of an inner conductor surrounded by a low loss insulating jacket and a braided copper outer conductor. The cable is covered by a waterproof vinyl jacket extruded over the braid. The inner conductor transmits the radio energy from one point to another and the outer conductor prevents the energy from escaping from the cable.

incoming signal before it arrives at your receiver. In short, a poor feedline severely cripples your CB station, both on receiving and transmitting!

The Coaxial Cable -- What Is It?

The *coaxial cable* (or concentric cable) is the most popular feedline in use for CB radio today. Over 500 different types of coaxial cable are manufactured, from microscopic, hair-like feedline for special instrumentation to giant line, several feet in diameter for super-power transmitters.

Coaxial line was originally designed at the turn of the century for use as submarine telegraph cables, and was modified for radio use in the early "thirties". Today's excellent, low-loss line was developed during World War II, and millions of miles of "coax" have been manufactured and used in the last 25 years.

An industry-wide standard for coaxial cable has been adopted, and most lines suitable for radio use are designated by the initials *RG* (derived from "radio guide"), followed by a serial number identifying a particular type of cable, and the suffix letter *U* (indicating "utility" cable). By far the most popular coaxial cables for CB operation are the RG-8/U and RG-58/U families of cable.

The flexible coaxial line (Figure 1) consists of a solid or stranded copper inner conductor, a solid or perhaps white foamy continuous insulating jacket (usually polyethylene) and a braided copper outer conductor. The outer conductor is covered by a tough, waterproof vinyl jacket extruded over the braid. The purpose of the inner conductor is to transmit the radio energy from one point to another and the braided outer conductor (shield) prevents radio energy from escaping from the line. When properly used, no radio energy exists on the *outside* surface of the line which may, in fact, be buried beneath the earth without affecting antenna performance. The insulating jacket protects the coaxial line and prevents it from being accidentally shorted to ground or to other wiring.

The coaxial line is flexible enough to be bent around a radius greater than six inches or so. Sharp bends should be avoided, as the center conductor can "cold flow" or slowly move about within the insulation when under continuous stress, and may gradually shift over and short itself out to the braid.

The Black Jacket

No, the Black Jacket is not a Japanese Judo Society, but the outer insulation of the coaxial line. Two types of jackets are used, and you should steer clear of one. The jackets are *regular vinyl* and *noncontaminating vinyl*. Older cables (made up to about 1965, and a few inexpensive cables still made today) use regular vinyl jackets which seem to have a chemical affinity for the polyethylene inner insulation of the line. Result: after a period of four or five years of life (and sooner at high temperatures) the radio frequency loss of the cable increases markedly and it becomes almost useless as a "radio hose".

How do you tell "lossy" cable? You can't unless you are well equipped with laboratory instruments, or can conduct a simple measurement as described later in this Handbook. The recommendation is: Don't buy surplus coaxial line, or cheap line made by an unknown manufacturer. The military surplus line has been sold by Uncle Sam because its useful life has been exceeded. Lossy coaxial cable is a perfect way to throw your costly transmitter power away!

Coaxial Cable Loss

Cable loss is measured in decibels, and may be thought of as "negative gain". As an example, suppose you have an antenna with a power gain of 3 decibels, and you have a coaxial line running to the antenna that has a power loss of 5 decibels. Your overall power gain is negative, and you end up with 3 - 5 = minus 2 decibels power loss!

Cable loss is directly proportional to the length of cable you use, all other factors being equal. You should therefore strive to have as short a run of coaxial cable between your CB equipment and the antenna as possible.

CABLE TYPE	DIAM.	ATTENUATION DB/100 FT.	NOMINAL IMPEDANCE (*OHMS*)	REMARKS
RG-8/U	.405	1.0	52	*OFTEN HAS CONTAMINATING JACKET*
RG-8A/U	.415	1.0	52	*NON-CONTAMINATING JACKET*
RG-213/U	.410	1.0	50	*NON-CONTAMINATING JACKET*
RG-58/U	.195	1.9	53.5	*OFTEN HAS CONTAMINATING JACKET*
RG-58A/U	.195	1.9	50	*TINNED COPPER CONDUCTOR*
RG-58C/U	.195	1.9	50	*EQUIVALENT TO RG-58A/U*

Fig. 2 COAXIAL CABLES recommended for CB use. RG-8/U and RG-8A/U are large diameter 52 ohm cables used for high power or long cable runs. RG-213/U is a newer version of RG-8A/U, standardized on a nominal impedance of 50 ohms. Older types of RG-8/U line have short life, caused by contamination of outer jacket; newer cables have long life jackets with a life expectancy in excess of fifteen years. For short cable runs and ease of installation, the small diameter type RG-58/U, RG-58A/U and RG-58C/U are recommended. The newer cable versions are standardized on a nominal impedance of 50 ohms. Many other lesser known 50 ohm cable types exist, but those listed are the most popular ones and the least expensive to buy.

The Noncontaminating Jacket

In case you are buying coaxial cable, you should buy the newer type that has the noncontaminating jacket that does not affect cable life. The new jacket style designation for RG-8/U is RG-8A/U. An even newer cable, of the same type with a noncontaminating jacket is RG-213/U, which has the same characteristics as RG-8A/U. The new jacket style designation for RG-58/U cable is RG-58A/U, or RG-58C/U. These cables are all tabulated in the chart of Figure 2.

The newer coaxial cables with a noncontaminating vinyl jacket have a life expectancy in excess of fifteen years. Considering that the extra cost of the improved cable runs only about a penny a foot, the noncontaminating types are a good investment for your CB station .

Foam Dielectric Cable

A recent addition to the family of coaxial cables is the *foamed dielectric* type, designed for use with cable TV (CATV). The inner insulation consists of cellular polyethylene, foamed with an inert gas. While well-suited for their original purpose, their use at CB frequencies is really not worth the extra cost as the improvement in efficiency is only a percent or so over less costly solid dielectric cables.

Your Choice of Cable

Figure 2 tells the story. Your CB equipment requires a 50 ohm transmission line. Actual coaxial line impedance runs from 50 ohms to nearly 54 ohms, but this manufacturing variation makes little difference in actual use. You are concerned with cable efficiency and long life. Vetoed are the cables with contaminating jackets, and those older cables with higher attenuation than others. The best group of cables to choose from are RG-8A/U and RG-213/U in the large diameter types and RG-58C/U in the small diameter type. At CB frequencies, the efficiency of a 100 foot run of the larger diameter cable is about 80 percent, and the efficiency of the smaller diameter cable is about 65 percent. Figure 2 shows you cable efficiency for various cable lengths. The moral is obvious: keep your transmission line as short as possible.

Line Impedance

Most popular coaxial lines are designated as "50 ohms" or "70 ohms", or some such cryptic value. Often, this value is embossed into the jacket of the line. What does it mean? This reference does not refer to the actual resistance of the line in ohms (which is very low) but to the *characteristic impedance* of the line, which is an electrical description of the line in terms of the *ratio of line voltage to line current*. (As a refresher in this topic, better look once again at Chapter 3 of this Handbook). As before, the term impedance refers to a *ratio* and not to a numerical value of resistance. The characteristic impedance ratio is determined only by the physical characteristics of the line: the inner conductor diameter, type of insulating material, and the thickness and construction of the outer shield. By varying the ratio of the diameter of the inner conductor to the shield conductor and changing the inner insulation, lines of 50 ohm impedance to 90 ohm impedance may be manufactured.

The *resistance* of the line, on the other hand, is altogether a different matter. While complex laboratory instruments are required to measure the characteristic impedance, only an ohmmeter is needed to measure the line resistance. For a short section of good coaxial line, the resistance of either the center conductor or the shield (as measured from one end of the line to the other) will be a fraction of an ohm.

What is the Correct Line Length?

How long should your coaxial line be? Is there a "magic" length which will make your antenna work better? Generally speaking, the answer to this question is: no. The line should be long enough to reach from the CB station to the antenna, and no longer. The shorter the line, on the other hand, the less will be the line loss.

Two exceptions to this general statement should be noted: First, some CB gear exhibits erratic tuning when operated into a transmission line having a high value of SWR on it. The solution is to lower the SWR by making proper tuning adjustments to the antenna or to alter the length of the transmission line a foot or so so that the ratio of voltage to current at the transmitter end of the line is more in accord with the limitations imposed by the output network of the equipment.

Second, some beam antennas use short lengths of coaxial line as coupling devices between various portions of the antenna. In this case, the antenna manufacturer usually specifies a particular line length to be used. It is wise to follow his instructions. Aside from these two special cases, follow Abraham Lincoln's answer to the question, "How long should a man's legs be?" Honest Abe replied, "Long enough to reach the ground."

Is Coaxial Line Waterproof?

Yes! The line itself is waterproof as long as the tough outer vinyl jacket is not cut. *But the ends of the coaxial line are not waterproof* and--contrary to popular wisdom--*water can easily get inside your coaxial line at the exposed end.* If it does, you are in trouble! The water will be sucked along the outer copper braid by capillary action until finally the whole line becomes soaked inside. The copper braid gradually corrodes, ruining line efficiency and reducing the capability of your CB station. This capillary action is very nerve-wracking and unless you live in Phoenix, Arizona (where it rarely rains), the open end of your coaxial line can suck moisture out of the air over a period of time, just as a small child sucks a cherry soda through a straw!

Remember, then, when using any coaxial transmission line you *must* waterproof the ends of the line to prevent moisture from entering.

How To Waterproof Your Coaxial Cable

It's a cinch to keep your transmission line dry on the inside. Just don't let water get into it in the first place by properly sealing the ends of the line. The *wrong way* to terminate your line is to peel back the waterproof jacket and make a connection to your antenna using the inner conductor and shield as wire leads wrapped around the antenna terminals. "Shucks", says Johnny Knowitall, "Why should I spend 70 cents for a coaxial connector? I can just twist the wires around the antenna bolts, and it works great!"

Poor Johnny. Moisture will condense at the end of his line when the air temperature drops and water will gradually work its way down the outer braid of the line, inch by inch. Rain water enters the coaxial line directly. Soon, *alle ist kaput*.

(A note from one author: I had this happen to me once, when I lived near New York City. I'd had an unsealed coaxial line up for a few years and took it down one day to move the antenna. When I dropped the end of the line on the ground, water ran out of it. It hadn't rained for weeks, and I couldn't imagine where the water came from. Then I learned that coaxial line can act like a water pipe! I drained nearly a glass of water out of the line. When I cut the line open, the copper braid was corroded, and a nasty green color. The inner insulation was discolored too. That taught me a lesson--and an expensive one, as I had to replace all the lines from the house to the antenna.)

Well, to get back to the subject: If you have a coaxial plug on the end of your line it is necessary to coat the plug and the receptacle with a waterproofing silicone sealant, such as General Electric RTV-102. This flexible silicone rubber comes in a ready-to-use tube, and adheres to almost any surface. It can be bought in white or translucent shades (either is OK) and dries to a tough, flexible coating that resists moisture and rot. The silicone is applied over the coaxial plug and receptacle after connection is tightly made, and allowed to dry for a few hours. When still "sticky", it is overwrapped with black, vinyl tape, such as *Scotch plastic electrical tape No. 88*. Application of the silicone compound, plus a double layer of tape, carefully wrapped will do the job.

If your line is terminated in leads instead of a coaxial plug, the silicone sealant should be carefully spread over the base of the leads

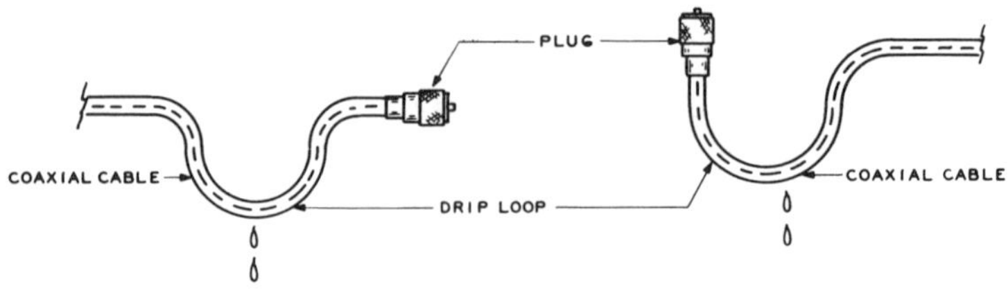

Fig. 3 COAXIAL CABLE IS WATERPROOF but most inexpensive fittings are not. In order to prevent water from entering the end of the line a drip loop should be used. In addition, the cable plug should be wrapped with vinyl tape and covered with a waterproofing compound as discussed in the text.

and pushed down into the end of the cable. The end of the cable itself is then carefully wrapped with vinyl tape. Use two layers of wrapping.

The Drip Loop

A *drip loop* such as shown in the illustration (Figure 3) keeps the water from flowing down the line into the coaxial fitting. Always adjust the placement of the cable so that water tends to run away from, and not towards, the coaxial fitting. Allowing water to run into a coaxial plug is asking for trouble, no matter how carefully you prepare the joint. Finally, tighten the coaxial fitting to the receptacle with pliers to make sure the outer ring of the plug established a good connection between the shield of the line and the frame of the receptacle.

Never splice two coaxial lines together without the use of the proper fittings. Properly prepared lines with UHF fittings may be spliced together with a minimum of fuss with the use of the double-ended coaxial fittings listed in Figure 4.

The Coaxial Plug: Your Best Friend

A bewildering collection of coaxial plugs is manufactured for all types of coaxial cable. By far the most popular are the so-called "UHF fittings", shown in the illustration. Most CB equipment uses these connectors. They are not too expensive, moderately easy to install

COAXIAL CONNECTOR TABLE		
"UHF" CONNECTORS FOR RG-8/U, RG-8A/U AND RG-213/U LINES		
DESCRIPTION	MILITARY TYPE	AMPHENOL TYPE
PLUG	PL259 PL259A	83-1SP 83-1SPA
SOLDERLESS PLUG	— —	83-151
SPLICE	PL-258	83-1J
REDUCTION ADAPTER FOR RG-58/U	UG-175/U	83-185
"BNC" CONNECTORS FOR RG-58/U, RG-58A/U AND RG-58C/U LINES		
DESCRIPTION	MILITARY TYPE	AMPHENOL TYPE
PLUG	UG-88/U UG-88B/U UG-88C/U	31-002 31-018 31-202
SPLICE	UG-914/U	31-219
ADAPTER TO UHF	UG-273/U	31-028

Fig. 4 COAXIAL PLUG, splice and adapters make life easy for you and give you a ship-shape antenna system when they are properly used. This table lists both military and Amphenol type numbers of most popular coaxial hardware. Other types and styles of fittings exist and long lists of special coaxial plugs and adapters are given in the larger radio parts catalogs. Special connectors are also available to match UHF and BNC fittings to RCA and Motorola plugs.

(if you know what you are doing), but they are *not* waterproof. Figure 4 shows a listing of the most-used UHF and BNC fittings for both RG-8A/U (large) and RG-58A/U (small) 50 ohm cable.

No doubt many fingers have been burned and many swear words have defiled the atmosphere from the frustration of incorrectly installing coaxial connectors. You can put 'em on the cable the hard way, but here's the *easy way* to do it!

How to Use Coaxial Plugs -- The Easy Way

Make neat and efficient connections between your coaxial cable and your CB radio set! Many CBers look upon coaxial plugs as inventions of the devil, and they may be right. Tricky to place on the line, when improperly installed, the common coaxial plug can cause intermittent operation and possible damage to the transmitter portion of your CB radio.

Getting the coaxial plug properly fitted on the end of a coaxial cable may be a frustrating and time consuming task, in spite of all the instructions and drawings of a correct assembly shown in magazines and handbooks. These wishful directions usually don't work. In many cases, the CB'er simply gives up, jams the connector on the cable, leaving short whiskers of copper braid ready to short out the antenna circuit -- and an open invitation for failure.

Properly done, the job of placing a coaxial plug on the transmission line is not difficult. The following assembly sequence was worked out over the years and is recommended for the popular type PL-259 coaxial plug, obtainable at any large radio supply house. This connector is intended for use with RG-8/U cable and (with the addition of an adapter) also with the smaller RG-58/U cable. You can determine if this style of plug fits your radio by looking at the antenna jack. If it is labelled SO-239 or 83-IR, it will match the PL-259.

The first step in preparing the coaxial cable for the plug is to slide the coupling ring of the plug onto the cable with the ring threads towards the open end of the cable. Next, take a utility knife (*Stanley 99A Shop Knife*, for example) and circumscribe a cut in the outer black jacket of the line. Make the cut at right angles to the cable and about 1½ inches back from the end of the line. The small cylinder of jacket material you have cut may be slit carefully with the knife and removed.

You have now exposed over an inch of the outer copper braid of the cable. Without disturbing the braid, which should be lying flat against the inner insulation, take a soldering gun and quickly and smoothly tin the exposed braid, making it a solid entity. Don't overheat the braid, or the inner insulation may melt and "squirt" out between the basket-weave strands of the braid. One or two practice runs on scrap cable ends will make an expert of you! Clean the left-over flux from the braid with a rag moistened with paint thinner or alcohol (See Figure 5).

The next step is to trim the soldered braid to the correct length. Use a small tubing cutter for this job. The *General Hardware #123 Midget Tubing Cutter* is recommended. Cut the braid so that 7/16 inch is left on the cable end. Mark a line this distance from the black vinyl jacket and place the tubing cutter over the braid, letting the cutting wheel fall on the mark. Tighten the cutter slightly and slowly revolve it about the cable. After one turn, tighten the wheel again, and continue to revolve the cutter. Four or five turns and the cutter will neatly slice the soldered braid. The unwanted slug of braid may be then snipped off with a pair of wire cutters. In using the cutter, don't cut down too far, or you'll slice into the center insulating material.

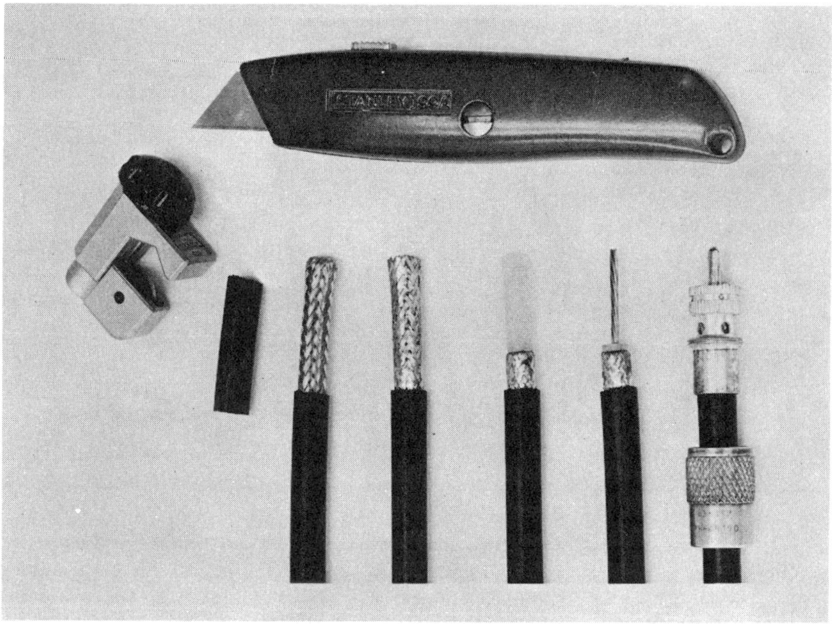

Fig. 5 COAX PLUG ONTO LINE--THE EASY WAY. This photograph shows the easy steps in preparing RG-8A/U coaxial line for a PL-259 coaxial plug. Midget tubing cutter (left) and utility knife (top) are used. At left is sample coaxial line with the outer jacket removed by the knife. Next, the outer, braided conductor of the line is tinned. Third view shows the outer braid cut to proper length by the tubing cutter. Fourth view shows the inner insulation of the line cut to length and inner conductor tinned. Right view shows coaxial plug and ring on line, with plug in position for soldering to line. Soldering is done through four holes in shell of plug. Soldering gun or iron with small tip and high wattage rating is recommended for this operation. Once the plug is soldered, it may be wiped with damp cloth to remove resin and to cool it.

The next step is to trim the center insulation. Cut it cleanly with the utility knife so that a collar 1/16 inch wide extends beyond the soldered braid. Easy! Don't nick the center conductor. Once the insulation is cut, you can pull the slug off the end of the cable by grasping it with your fingers and gently pulling it, rotating it at the same time so that it follows the twist of the inner conductor wires. When the slug is off, tin the center conductor.

Now the cable end is ready for the PL-259 plug. Push it carefully on the cable end, rotating it with the fingers so that the internal threads of the plug screw onto the outer vinyl jacket of the cable. Make sure the inner conductor is centered into the plug pin. As the plug body is

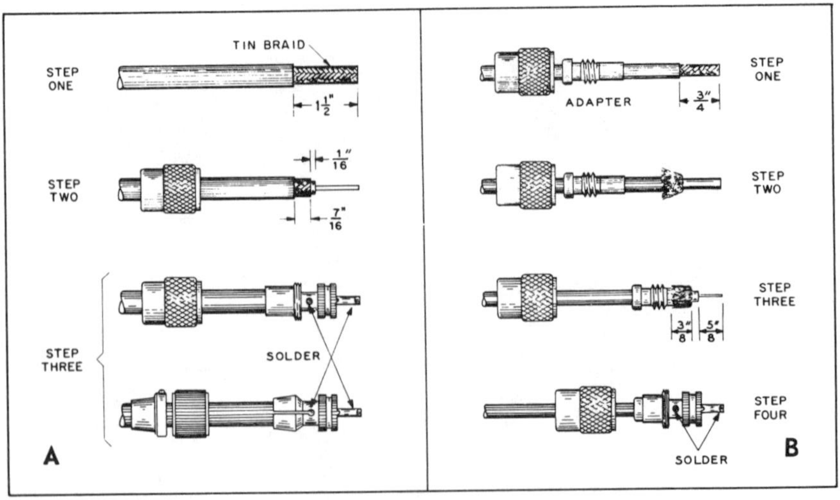

Fig. 6 UHF PLUG AND ADAPTER installed on a coaxial line -- the easy way. PL-259A plug installation is covered in (A), corresponding to the steps shown in Fig. 5. PL-259 and UG-175/U adapter are shown in (B) matching the small diameter RG-58A/U coaxial line to the large diameter coaxial plug.

screwed onto the cable, you'll see the tinned braid appear through the four solder holes in the shell. Continue twisting the plug onto the cable until the braid is completely visible through all holes (Figure 6A).

Now is the time to solder the plug onto the cable, the idea being to solder the braid through the four solder holes. Use a soldering gun with a small tip and proceed with care, using small diameter solder. It is usually easier to hold the plug in a bench vise during this operation. Take care that solder does not run over the outer threads of the body. The last step is to solder the center conductor to the plug pin. After the assembly cools down, slide the coupling ring down over the plug.

Try this technique on a plug and a spare piece of coaxial line. Soon you'll be able to produce masterpieces that will arouse the envy of your friends. A real artist, once he masters this simple technique, can then advance to the reverse step -- salvaging coaxial plugs from old cable!

The PL-259 Plug with Small Coaxial Cable

The PL-259 may be used with small diameter cable (RG-58/U, for example) by adding a reduction adapter (type UG-175/U). A slightly different assembly technique is used for this cable (Figure 6B).

The end of the cable is passed through the coupling ring and the adapter, with the threads of the ring and the narrow end of the adapter facing the end of the cable. Using the utility knife, cut 3/4 inch of the vinyl jacket off the cable. Fan the braid out slightly and carefully fold it back over the adapter.

Next, trim the braid with a small scissors to about 3/8 inch long, so that it fits about the barrel of the adapter. Following this, take the utility knife and remove 5/8 inch of the insulation from the center conductor. Careful! Don't nick the conductor. Finally, tin the conductor.

Now, carefully screw the plug body onto the adapter. The center conductor of the cable should pass easily through the center pin of the plug, and the strands of the braid should appear through the side holes of the shell. Using a small soldering gun, solder the braid through the holes. Lastly, solder the center conductor to the plug pin and slide the coupling ring down over the plug.

Check Your Coaxial Line!

The first step to take when antenna trouble is suspected is to carefully check your coaxial line, especially the plugs at both ends. By far the greatest number of antenna difficulties stem from incorrectly soldered coaxial line fittings. Then, too, in checking an old feedline installation, or checking someone else's line, the end fittings should be carefully examined.

You can perform an important check on your line with the aid of an inexpensive volt-ohmmeter (Figure 7). The ohmmeter should be set on the *highest resistance* range and the coaxial line disconnected at both ends. By measuring the resistance between outer braid and inner conductor at one end of the line you can make sure that a leakage path does not exist *across* the line from braid to inner conductor. The resistance reading between inner and outer conductor on a good piece of coaxial cable should be *infinity*. If a resistance value of a few hundred ohms (or several thousand ohms, for that matter) is noted, the coaxial fittings should be immediately suspect. In addition, the line itself should be closely examined for abrasions, breaks in the jacket, and damage to the copper braid under the jacket. If a coaxial plug is in doubt, it should be cut off, discarded, and a new one substituted in its place.

Finally, using the ohmmeter on the *lowest resistance* range, measure the cable from one end to the other along the same conductor. Both the outer braid and inner conductor should show an end-to-end resistance of less than one ohm or so. Thus, a good coaxial line shows infinite re-

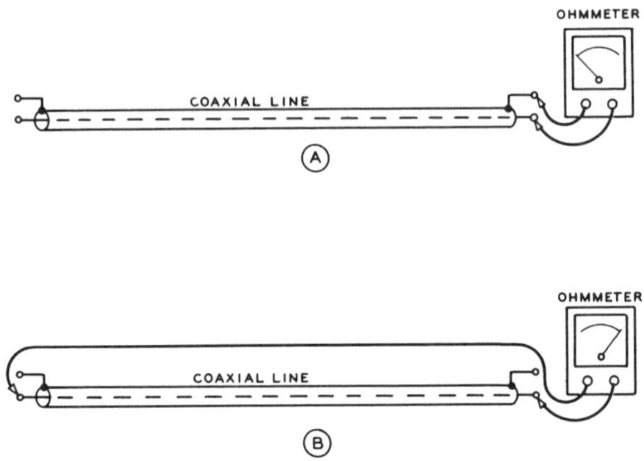

Fig. 7 TEST YOUR COAXIAL LINE with an ohmmeter. Leakage across line is measured between conductors (A) with meter on highest scale. Resistance of each conductor is measured along conductor (B) with meter on lowest scale,

sistance *between* the conductors and near-zero resistance *along* each conductor from end to end.

The final coaxial line check is to run transmitter power through the line into a *dummy load* and see how things perform. Place the dummy load at the far end of the line, hook the opposite end of the line to your CB radio through an *SWR meter*. Tune up the equipment in the normal manner. A good coax line will show practically no reverse reading on the SWR meter after the instrument is set for a forward, full-scale reading (in other words, the SWR is unity, or 1). If the coaxial line passes this test it is A-Okay and will deliver power to your antenna. There's more on the SWR meter in the next chapter, and dummy loads are discussed in Chapter 14.

Line Loss Vs. SWR

The line attenuation given in Figure 2 assumes that the SWR on the transmission line is unity. What happens to line loss at high values of

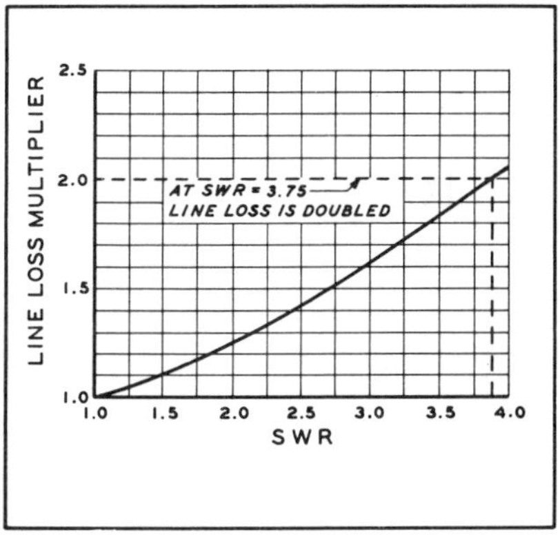

Fig. 8 COAXIAL LINE LOSS increases as SWR increases. Line loss will be doubled when SWR is 3.7. Reducing SWR drops line loss. Lowest loss is achieved by using large diameter coaxial line such as RG-8A/U.

SWR? The answer is that line loss increases with increasing values of SWR as shown in Figure 8. For example, if a transmission line has an SWR value of 3.7, the line loss will be doubled. A 100 foot length of RG-58A/U, for example, has a nominal line loss of 1.9 decibel. If the SWR is 3.7, the line loss increases to 3.8 decibels. This corresponds to a 60 percent loss of power! Changing from RG-58A/U to RG-8A/U will drop the line loss from 3.7 decibels down to 2 decibels, which corresponds to a 38 percent loss of power.

For long runs of transmission line, it is prudent to use the heavier and more expensive RG-8A/U line rather than the smaller RG-58A/U cable. In any event, the overall length of the transmission line should be held as short as possible.

Chapter 10

The SWR Meter

(Friend or Foe?)

How can you be sure your CB antenna is operating at its highest degree of efficiency? Unlucky Pierre, the well-known CB operator says, "Yes! My antenna is perfect, but I cannot talk to anyone."

What's wrong with Unlucky Pierre's antenna? Is he merely the victim of bad luck and interference on the channel, or has he made a monumental "goof" in assembling his antenna that has rendered his expensive station useless?

In a few moments Pierre can get an answer to his antenna problems--which may be real, or merely imaginary--by making simple measurements with an instrument known variously as an *SWR meter, reflectometer,* or *SWR bridge* (Figure 1). This device (placed in the transmission line) reads the *standing wave ratio (SWR)* of the antenna system on a meter and by measurement of this ratio tells Pierre how well his antenna system is working. It will tell you the same thing, too!

Incident And Reflected Waves

Radio waves travelling through space, or along a transmission line, have been compared to water waves made by a stone thrown into a quiet pond. The waves travel outward from the stone in expanding, concentric circles until they meet an obstacle, at which point they are reflected back towards the source. The forward travelling water waves *(incident waves)* and the *reflected waves* interact with each other and form interesting patterns of interference on the surface of the water.

Another fanciful analogy to radio waves is of a person holding a long rope attached to a post. He can start flipping the end of the rope, creating a wave of motion moving along the rope towards the post. Upon reaching the post, the wave will be reflected back towards the opposite

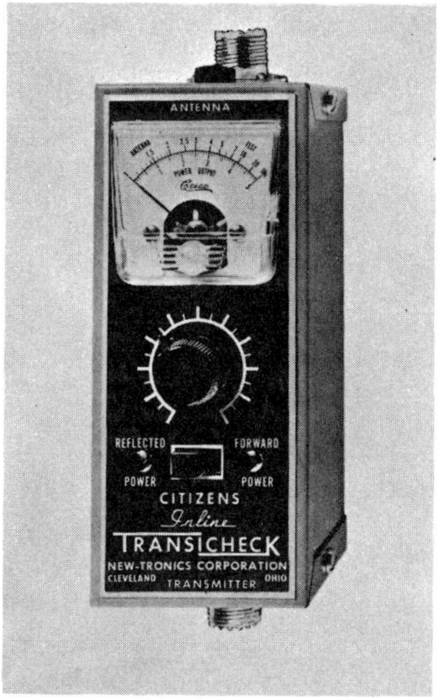

Fig. 1 THE SWR METER is useful instrument for your CB station. It measures the ratio of the incident (forward) to the reflected power in your transmission line. This ratio is termed the standing wave ratio. The SWR meter is placed in series with your transmission line between the CB transmitter and the antenna. Power is applied from the transmitter and the potentiometer control is set for a full scale meter reading when the control switch is in the "forward power" position. Switch is then thrown to "reverse power" position and the standing wave ratio is read directly from the meter of the instrument.

end. If a number of equally timed flips are given, a succession of waves at equal intervals are sent along the rope. When reflected back from the post at the far end, they meet others coming along whose "wavelength" is equal to that of the waves coming back. At some points the rope tends to move a certain distance upwards with the direct wave, and the same distance in the downwards direction with the reflected wave; the result is that at these points the rope does not seem to move at all. These points are found along the rope one half wavelength apart. At all other points the rope moves (or vibrates) in the resultant direction of the combined direct and reflected wave impulses, creating what are called *stationary waves,* or *standing waves.* Waves of this type can be set up along a conductor by suitably timed electrical impulses applied to the conductor. A picture of the standing wave analogy on a rope is shown in Figure 4, Chapter 2.

What Is The Standing Wave Ratio (SWR)?

A combination of incident and reflected electric waves on a radio conductor will result in a combined wave called a *standing wave.* The ratio of size (amplitude) of the incident to the reflected wave is termed the *standing wave ratio* (abbreviated SWR). The standing wave ratio may be measured by a suitable instrument.

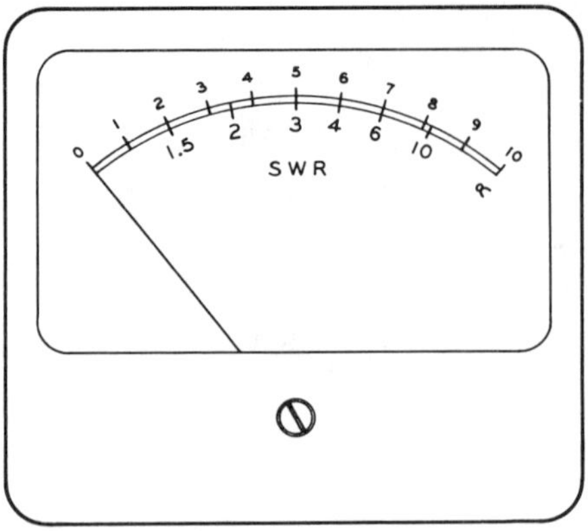

Fig. 2 TYPICAL SWR METER CALIBRATION. The top scale is merely for reference. Bottom scale is calibrated in SWR. Lowest SWR reading is at left of scale. SWR readings above 10 are usually not accurate and most meters are calibrated only to that value. SWR meter measures forward and reverse power in transmission line, but scale is calibrated in terms of SWR for easier interpretation. You can buy a good SWR meter for about ten dollars.

The SWR on a coaxial transmission line is measured by an SWR meter and the magnitude of the SWR is read directly from the scale of the instrument. The SWR is expressed as a ratio, with unity (zero reading) indicating no reflected power and infinity (full scale reading) indicating a state of maximum reflection. A typical SWR meter scale is calibrated as shown in Figure 2.

Since the electrical impulses consist of voltage and current waves which follow each other regularly, the standing wave as a result, is composed of waves of voltage and current. As a definite relationship exists between the voltage and current waves, the amplitude, or strength of the standing wave may be ascertained by measuring either the voltage or the current in the proper manner. Most inexpensive SWR meters monitor the voltages in a transmission line and compare the incident voltage with the reflected voltage.

With a perfect antenna system, the reflected wave is zero and SWR is termed to be *one, unity,* or *one-to-one.* As the reflected wave increases in amplitude, the SWR increases. The SWR reading, in effect, shows the *degree of mismatch* between the impedance of the transmission line and that of the antenna itself, as discussed in the following section.

Antenna Matching

When the radiation resistance of the antenna (that realistic voltage to current ratio) is the same value as the characteristic impedance of the transmission line (the ratio of voltage to current in the line), all of the electric energy transmitted down the line is taken by the antenna and radiated into space in the form of a radio wave. If the antenna does not match the transmission line perfectly, wave interference is set up at the junction of the line and the antenna and a portion of the energy is reflected by this mismatch point back down the transmission line towards the transmitter. As the degree of mismatch between the antenna and the line increases, the amount of reflected radio energy increases, just as a larger obstacle tossed in the water pond will reflect back a bigger wave of water. Standing waves on the transmission line are thus created by the interaction of the forward and reverse travelling electric waves.

As the reflected wave increases, the incident wave decreases, but the total power in the two waves remains the same. The worst possible case exists when the incident and reflected waves are equal in strength. This happens when the far end of the transmission line (the end away from the transmitter) is either short circuited or open.

The SWR Meter--How Does It Work?

With the idea of incident and reflected waves under our belt, it's not too difficult to understand the operation of the SWR meter, a schematic of which is shown in Figure 3. This version of the SWR meter (and there are others) takes advantage of the fact that the voltage and current of the forward wave are in step (in phase) while the voltage and current of the reflected wave are out of step (out of phase).

The SWR meter is placed in the coaxial line. It has two small pickup wires placed near the inner conductor of a section of line. Current is coupled into a pickup wire when a radio wave passes down the line and a very small amount of voltage is sampled through the capacity of the wire to the line.

The current in the pickup wire flows through a resistor at the end of the wire and the voltage across the resistor is measured by a sensitive meter. Two pickup wires are incorporated in the SWR meter, reversed with respect to each other. When the SWR on the line is low, the pickup wire voltages and currents are in step and are low in value. As the SWR increases on the line, the pickup wire voltages and currents are not in step, and are higher in value. The voltmeter consequently reads a resultant voltage. Thus, by using two pickup lines, reversed with respect to each other, currents and voltages flowing in either direction in the line may be compared. Since the SWR meter is symmetrical, either pickup unit may be used for either forward or reflected measurements and both readings may be applied to one meter via a selector switch.

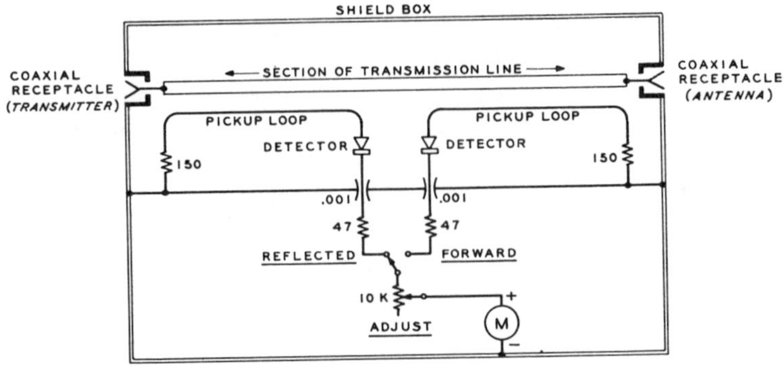

Fig. 3 SCHEMATIC OF TYPICAL SWR METER. Many different types of SWR meters exist, but this version is commonly used in inexpensive designs. The sensing device is a short section of transmission line inclosed in a shielded box. Twin pickup loops sense the voltage and current flowing in the incident and reflected waves. Selector switch permits operator to observe relative voltages on meter of instrument.

Using The SWR Meter

The comings and goings of the radio power in your transmission line can be neatly separated by the SWR meter and the power travelling in each direction can be read on the meter of the instrument. Adjustments may be made to the antenna in order to reduce the reverse (or reflected) meter reading, thus dropping the value of the SWR on the transmission line. *The name of the game is to reduce the reverse reading to as low a value as possible,* say, less than 10 percent of the forward reading, or to as near-zero as possible. Making such antenna adjustments without the use of an SWR meter is like washing your feet with your socks on-- you can do it, but it ain't easy!

The SWR Meter As A Tuning Aid

Why bother with an SWR meter? Well, the SWR meter is a powerful tool to aid you in obtaining optimum antenna performance. At the same time it may be used to tune your CB transmitter for maximum power output. Through maladjustment, it is possible to drop transmitter power output by as much as 90 percent, without the operator being aware of the loss of power! The proper use of the SWR meter will eliminate catastrophies of this type.

In addition to giving the operator a tool to determine proper transmitter tuning, the SWR meter provides a handy means of judging antenna adjustment. If the SWR on the transmission line to the antenna is high, it indicates a bad degree of mismatch between the antenna and the transmission line. *No adjustment or tuning done at the station can cure this difficulty,* only adjustments made to the antenna can possibly lower the SWR on the transmission line. Incorrect antenna adjustment causes a high value of SWR and this, in turn, causes an increase in power lost in the line and a drop in efficiency of your transmitter. Indeed, some "solid state" (completely transistorized) CB transmitters will not operate into a transmission line having a high value of SWR and--in extreme cases--damage may result to the rf power transistor in the equipment as a result of high SWR on the transmission line.

A Trial Run With Your SWR Meter

Let's make an actual SWR measurement with an inexpensive SWR meter, such as shown in the photograph. The SWR meter should be placed in series with your transmission line, near your operating position, so that a check may be made on both transmitter operation and SWR meter readings. Most inexpensive SWR meters are equipped with SO-239 type coaxial receptacles which match the common PL-259 type coaxial plug. Connect the output (antenna) receptacle of the SWR meter to your transmission line and antenna, and connect the input (transmitter) receptacle of the SWR meter to your CB set through a short, 50 ohm coaxial line about two feet long, or so. One end of this line should have a PL-259 matching plug on it, and a plug suitable for your CB radio is placed on the opposite end. Since you, good reader, have just finished reading the section of this Handbook on the fine art of placing plugs on coaxial lines, we'll assume that your coaxial connections are properly made! Your complete installation is pictured in Figure 4.

Make sure that the coaxial line used between SWR meter and transmitter is the same type as the line running to the antenna. By far the greatest number of CB radios and CB antennas are designed for 50 ohm line, and most SWR meters are designed for 50 ohm line, too. *However,* 70 ohm line does exist, and some SWR meters are made for 70 ohm line. Don't mix up a 70 ohm SWR meter with a 50 ohm line, or you'll get funny readings that don't mean a thing! All bets are off, too, if you

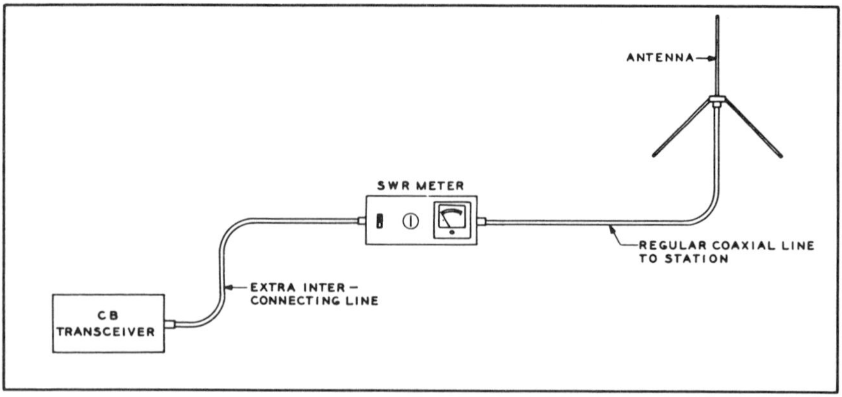

Fig. 4 TYPICAL SWR METER INSTALLATION. The instrument is placed in series with your coaxial line near the operating position. A short, extra length of coaxial line runs from the CB equipment to the SWR meter. The instrument may be left permanently in the line as a check on transmitter and antenna operation. Sensitivity control on panel of SWR meter permits use with transmitter powers up to several hundred watts.

try and make-do without the use of coaxial connectors, as you'll get screwy readings when the radio energy runs all over the place. Keep the energy *inside* the line where it belongs!

Once the connections to the SWR meter are properly made, the panel transfer switch of the device is set for a *forward* (FWD) reading, and the *sensitivity* control is set to minimum sensitivity (maximum counter-clockwise position). Turn on your CB transmitter and tune it up in the normal manner. Now, advance the sensitivity control on the SWR meter for full scale deflection with the transfer switch set in the *forward* position. Without touching the sensitivity control setting, now throw the switch to the *reverse* position and note the meter reading. If all is well, the reverse reading will be substantially less than the forward reading. Most SWR meters have a scale calibrated in "Standing Wave Ratio" and the reverse reading is taken from this scale. If all is well with your antenna system the reverse reading will be less than 2. A really well adjusted (or "matched") antenna will present a reverse reading of less than 1.5 or so. A perfect match will result in a reverse reading of zero.

If your reverse reading is over 2.5 (the mini-whip excepted), or is nearly as high as the forward reading, you are probably in trouble. Either the instrument is faulty, incorrectly placed in the circuit, or the antenna or feedline is malfunctioning. A very high reverse reading usually indicates either an open connection or short circuit exists in the antenna system somewhere. See page 117 for SWR data on mini-whip.

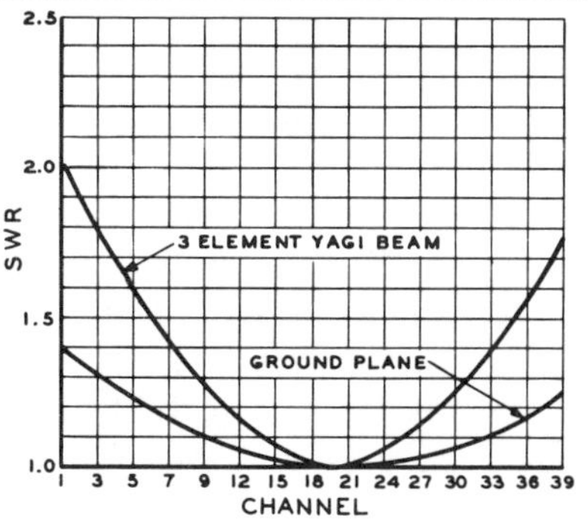

Fig. 5 TYPICAL SWR CURVES for Yagi beam and ground plane antenna. Chart is made up by noting SWR reading on various CB channels and plotting SWR versus channel or frequency.

You'll notice that when you run this test, the reverse SWR meter reading varies from channel to channel. This is normal, and you can actually plot a curve of SWR indication versus channel number, as shown in Figure 5. The curve should be smooth and the minimum SWR point should fall about the middle of the CB frequency range. Make a separate SWR reading for each channel, logging both forward and reverse readings. Plot the reverse (SWR) reading on a sheet of graph paper and you'll have a record of the performance of your antenna. Later, if you suspect antenna trouble, you can re-run the SWR measurements and compare them with the original curve.

Transmitter Tuning With The SWR Meter

The SWR meter may be used to peak the transmitter controls for maximum power output, or to check output. Merely tune your transmitter output stage for greatest meter reading with the SWR meter switch set in the *forward* position. The meter reading is not calibrated in watts, but merely gives greatest indication when the transmitter output is maximum. Make sure you do not drive the meter off-scale (retard the sensitivity control if necessary) or the instrument may be damaged.

Friend or Foe?

To repeat: the lower the reverse reading on the SWR meter as compared against the forward reading, the lower the SWR value of your antenna system. If the antenna match is perfect, the reverse reading will be zero. Most properly designed CB antennas exhibit a reverse SWR reading of 2, or less, across all CB channels.

The question might well be raised: just what is the maximum value of reverse SWR reading that may be encountered in a typical antenna installation while still assuming that the antenna is working properly? Generally speaking, if the SWR reading is greater than 2 or so, it may be an indication of improper operation and (other things being equal) possibly could indicate that things are amiss at the antenna end of the coaxial line. An SWR of 3 or more may be a definite indication of antenna or feedline malfunction. Good CB antennas hold the SWR across all channels to an indication of about 2, and SWR readings of this magnitude should be no cause for concern as the mismatch is trivial and may be ignored.

An exception to this statement is the special situation involving a very short antenna, such as the mobile mini-whip (see Chapter 11). The mini-whip characteristics are such as to reveal a high value of SWR on the transmission line when the whip is operated at a frequency removed from that to which it is tuned. SWR values of 3 or more may be observed with a mini-whip operated off-channel as discussed in the following chapter.

Finally, we come face-to-face with the question at the front of this chapter: SWR Meter--friend or foe? The answer is that the SWR meter can be the best friend you ever had *if* you use it properly and can be your worst enemy if you don't understand the significance of the readings. Take time to measure the SWR of your antenna system on all the channels you use. Make a graph of the readings, as shown in the illustration. Keep the graph and re-check your readings every month or so. If trouble develops in your antenna, you'll notice that the SWR readings have changed. You can continuously check antenna operation by leaving the SWR meter in the transmission line. It will indicate your transmitter output power, and a glance at the meter once in a while is a great satisfaction as you realize your equipment is operating properly.

The Field Strength Meter

Some SWR meters incorporate extra circuitry that permit them to be used as a *field strength meter*. This device measures the strength of the radio field in the vicinity of the antenna. It is possible to tune the CB transmitter for maximum field strength with the aid of this meter, but readings are severely affected by objects in the vicinity of the antenna, including people moving about. Used with caution, the field strength meter can tell you how your antenna is radiating.

Chapter 11

Your Mobile Antenna

(Poor? Better? Best?)

Because of space limitations on the vehicle, the "work horse" antenna for CB mobile service is the *vertical whip* shown in various versions in Figure 1. The most popular whip is a flexible, quarter wavelength tapered steel rod with a threaded base fitting. The top end of the rod should have a plastic ball on it to prevent you from putting your eye out when installing it on your car. In addition, the plastic ball insulates the tip and prevents static electricity from building up on the antenna, as often happens in a fast-moving vehicle in dry weather.

Whip antennas come in all sizes, price ranges, and performance values, and this chapter discusses some of the more popular and practical mobile antennas and their use.

The Quarter Wavelength Whip Antenna

The quarter wavelength whip antenna for the CB channels is about 105 inches long overall. Part of the antenna length is included in the mounting fixture and, as a result, manufactured whips take this into account and often run about 102 inches long. Length is not critical within two or three inches. The whip antenna requires no radials as it makes use of the metal frame of the vehicle as an artificial ground. The combination of vehicle and whip may be thought of as a distorted ground plane antenna. The fact that the car body is part of the antenna system can be readily ascertained during transmission by an SWR meter in the transmission line between the CB set in the car and the whip antenna.

Fig. 1 TAKE YOUR CHOICE of different mobile whip antennas for CB service. This typical young lady CB operator ponders which glamourous antenna she should put on her imported sports car. Mini-whip? Trunk mount? Obviously, she seems worried.

Opening the car doors, or standing on the ground and touching the car body with your hand, while transmitting will make a definite change in the SWR reading.

The quarter wavelength whip is an unwieldy antenna, capable of striking the top of your garage entrance, and long enough to bang into overhead tree limbs when you drive down the street. Nevertheless, it is the most efficient and simple mobile antenna available, and its use is recommended wherever possible in spite of its height.

The radiation resistance of the quarter wavelength whip antenna falls very close to 40 ohms and provides a reasonably good impedance match to a 50 ohm coaxial transmission line. The line, therefore, may be run directly from the CB equipment to the whip without the nuisance of placing a tricky matching device (antenna tuner) at the base of the whip to match the antenna to the line.

Some whip antennas are made of Fiberglas with a copper strap buried within the insulating material. These whips are rugged but expensive; they are recommended for use in areas where salt spray in the atmosphere, smog, or dampness may lead to rapid corrosion of the metal whip.

All in all, the quarter wavelength whip antenna is the best choice for CBers demanding good efficiency and low cost in their mobile antenna installation. When a compromise must be made, other antennas are available, as we shall see.

The Mini-Whip Antenna

Short, loaded miniature whip antennas may be purchased for 27 MHz CB operation. These whips are from 18 inches to 60 inches long. The missing antenna length that makes up the 102 inches required for a quarter wavelength resonant antenna takes the form of a *loading coil* placed in series with the whip. The coil may be placed either at the base of the whip, or near the center. Since a portion of the normal, full-size antenna is missing (it is wound up in the loading coil), the efficiency of the whip suffers to a degree. Short (18'' to 36'' or so) base loaded CB whips are about ten to twenty percent as efficient as a full size whip, with the greater portion of the transmitting power wasted in the form of heat in the loading coil. The longer mini-whips, of course, have the higher efficiency. It is unfortunate that the short whips do not perform better, but the efficiency of any antenna drops very sharply when it is less than one-quarter wavelength long and the missing length takes the form of a loading coil. The efficiency of the antenna may be expressed in terms of the *Q Factor* of the loading coil, as discussed next.

The Q Factor

The Q Factor (often spoken of as Q) is the figure of merit of any antenna or coil (inductance). A beam, dipole or quarter wavelength whip antenna is quite efficient and has a high figure of merit, or Q Factor. The Q Factor takes into account the overall efficiency of the whole antenna, including the loading coil used in mini-whips. Resonant antennas, such as those mentioned previously, use no loading coils while the proper operation of the mini-whip requires the addition of a loading coil to establish resonance. Alas, the loading coil is the weakest link in the chain of components that makes up the antenna system. To achieve a high Q Factor in the loading coil, it should be quite large, wound on a low-loss form, and use very heavy wire. For reasons of cost,

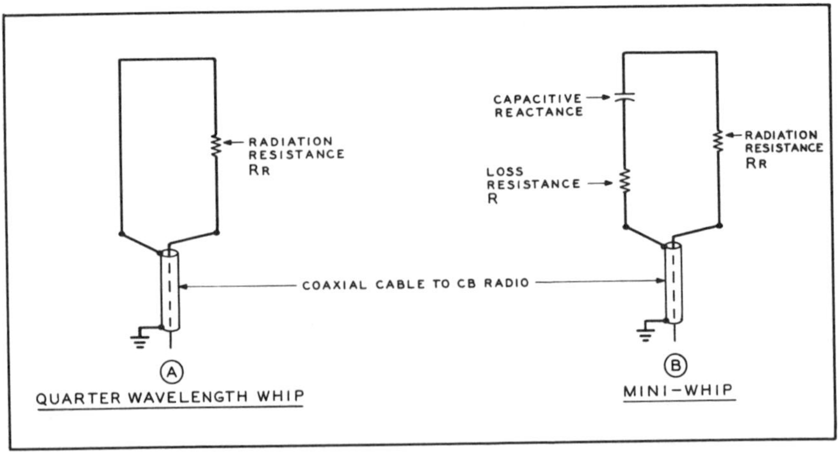

Fig. 2 ELECTRICALLY EQUIVALENT circuit of quarter wavelength whip antenna (A) and mini-whip antenna (B). Loss resistance of mini-whip wastes transmitter power and capacitive reactance must be neutralized by addition of loading coil to circuit.

eye-appeal and wind resistance, most loading coils are wound with small diameter wire on small, inexpensive, lossy coil forms. The resulting loading coil usually has a low Q Factor and is relatively inefficient. As a result, the mini-whip antenna using such a coil is inefficient, too.

A typical high Q Factor loading coil for a mini-whip antenna might be wound of #12 copper wire, perhaps one inch in diameter and 3 inches long. The number of turns in the coil depends, of course, upon the overall length of the mini-whip. Does *your* mini-whip loading coil resemble this description?

Radiation Resistance of a Mini-Whip Antenna

The radiation resistance of an antenna is proportional to the size of the antenna compared to the size of the radio wave. Large antennas, generally speaking, have high values of radiation resistance and small antennas have low values of radiation resistance. Practical, full size dipole antennas and beams have radiation resistance values ranging from about 20 ohms to 100 ohms. These values are "in the ball park" when a 50 ohm transmission line system is used, and result in SWR measurements of 2 or less on the transmission line, even without the use of a matching transformer between antenna and line. If such a transformer is used to lower the SWR, it is quite efficient, as the transformation ratio is not large.

The radiation resistance of a mini-whip, on the other hand, is a horse of a different color. The radiation resistance of a four foot long, coil loaded whip for the CB channels, for example, is about 5 ohms. The radiation resistance of a two foot long, coil loaded whip at the same frequencies is about 2 ohms. In order to efficiently couple the power from the transmitter and transmission line (which are designed for a nominal 50 ohm antenna load), some form of matching device must be used to make the system compatible with the very low radiation resistance of the mini-whip. To complicate matters, in addition to showing a very low value of radiation resistance, the mini-whip itself also exhibits a large value of *capacitive reactance* (the electrical characteristics of a capacitor, or condenser). The electrical circuit of a short mini-whip is shown in Figure 2. Since the radiation resistance is very low, a high value of current must flow in this circuit for a given power level. The current must flow through the large capacitive reactance, since it is electrically in series with the radiation resistance of the antenna. Unfortunately, little current will flow through this reactance as it is many times larger than the radiation resistance of the antenna. The reactance, therefore, must be cancelled out by the addition of an equivalent amount of *inductive reactance*. The loading coil supplies the inductive reactance and the mini-whip, plus the sum of the capacitive and inductive reactance form a resonant circuit at the operating frequency. This resonant circuit removes the reactances from the problem, but still leaves the very low radiation resistance of the mini-whip to be accounted for. By means of a suitable matching transformer placed at the base of the mini-whip, the antenna may be matched to the 50 ohm transmission line.

Unfortunately, as we have pointed out, all loading coils have a Q Factor, and some coils are quite inefficient and dissipate a large amount of your transmitter power as heat. This inefficiency may be thought of as a resistor placed in series with the antenna circuit. Adding the loading coil and matching transformer makes it possible for you to transfer power to the mini-whip, but you have paid for this privilege by accepting a loss in efficiency in the antenna system. This loss is determined mainly by the Q Factor of the loading coil and to a lesser degree by the Q Factor of the matching transformer. Even the best of loading coils is none too good. Time spent in choosing your loading coil for a mini-whip will pay big dividends in increased operating range of your mobile CB radio installation.

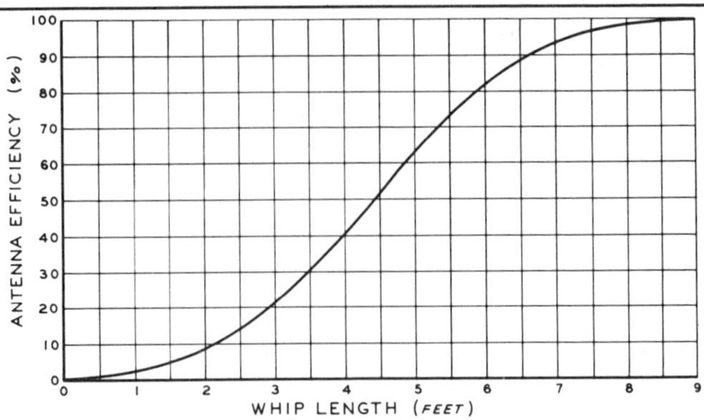

Fig. 3 MINI-WHIP antenna efficiency drops sharply as whip length decreases. CB whip antennas much shorter than four feet or so are very inefficient and user pays for compactness with a large drop in signal strength, both on transmitting and receiving.

Minimizing Antenna Losses

The graph of Figure 3 shows the approximate overall efficiency of a mini-whip and loading coil expressed as a function of antenna length. It is assumed that a coil with a high Q Factor is used. This generalized graph indicates that antenna efficiency drops sharply with whip lengths much less than about six feet, and that a very short CB whip -- even when a loading coil with a high Q Factor is used -- is relatively inefficient. Efficiency of a mini-whip using a loading coil with a low Q Factor can be much worse than the graph indicates. Unfortunately, Q Factor can only be measured by an expensive instrument called a *Q Meter*; but few of these instruments are available outside a communications laboratory. Even so, it is possible for you to estimate the relative Q Factor of a loading coil by examining the wire size, length, diameter and coil form material. Heavy wire, large diameter, spaced turns and a ceramic coil form are signs of high Q Factor. Look for these tell-tale signs before you buy your mini-whip antenna.

Center Loading

Theoretically, it is possible to place the loading coil at any point in the antenna. Placing the coil at the base of a mini-whip makes a physically strong assembly with low wind resistance. Antenna efficiency, however, can be raised an appreciable amount by placing the loading coil near the center of the antenna rather than at the base. Raising the coil beyond the center of the antenna does not "buy" much, and tends to make the antenna top heavy and adjustment more critical.

Many mini-whip antennas have the loading coil near the center point in order to boost antenna efficiency, and then cut down the length and diameter of the coil to decrease wind resistance. The improvement in efficiency in such circumstances is doubtful. Center loading, however, requires fewer turns in the loading coil for a given whip length than does base loading, and higher Q Factor is obtained as turns are removed from the loading coil. As stated before, the actual Q Factor is difficult to measure without the proper instrument and the wise CBer can only judge the merit of a mini-whip by the construction and placement of the very important and critical loading coil.

Operating Range of the Mini-Whip Antenna

In addition to lower efficiency, the coil loaded mini-whip is more difficult to adjust than is the full size whip antenna. In particular, the length of the mini-whip *above* the loading coil is quite critical. The overall length of a full size whip antenna is critical to within a few inches, whereas the length of the top portion of a mini-whip antenna is critical to one-half inch or less. Indeed, the better mini-whips have an adjustable tip section to permit the operator to vary whip length for optimum operation.

Proper whip length adjustment may be ascertained by the use of an SWR meter installed in the coaxial transmission line running from the CB equipment to the whip. The reverse SWR reading is checked on various channels. The full size whip antenna, properly adjusted as to length, will show an SWR curve that resembles curve A of Figure 4. A

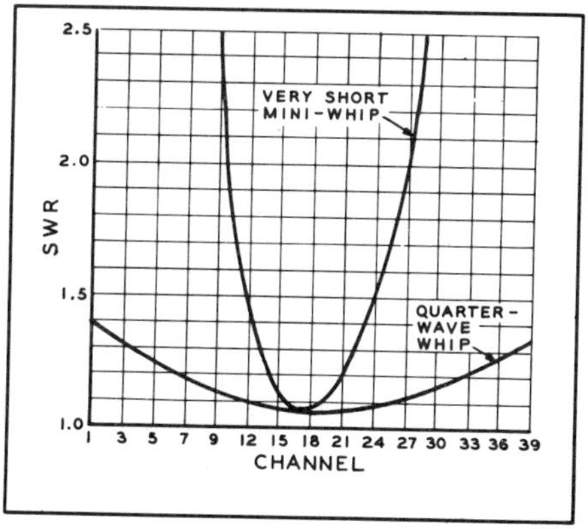

Fig. 4 TYPICAL SWR CURVE of mini-whip antenna is very sharp compared to the curve of full size whip. Mini-whip must be adjusted to channel in use or efficiency suffers and SWR may reach unreasonably high value on coax line.

typical mini-whip, properly adjusted, shows a resemblance to curve B. Each antenna in the illustration is tuned for operation at the approximate mid-frequency point of the CB channels. Note that the coil loaded mini-whip has a much steeper SWR curve and that the SWR at the higher and lower frequency CB channels is quite severe. This indicates that the mini-whip must be adjusted to the CB channel most frequently used in order to achieve optimum operation; operation on channels far removed from this channel must put up with a higher than usual value of SWR on the transmission line.

A Quick Check of Antenna Adjustment

Any whip antenna "sees" a different electrical environment at different locations on the vehicle. Since the metal area around the whip serves as a kind of ground plane, its effect will vary with antenna location and the profile of the vehicle.

If the electrical "shape" of the vehicle's body and the placement of the whip antenna are not optimum, a high value of SWR may exist on the transmission line to the antenna. For this reason, many antennas require "pruning" in order to reduce the SWR reading to a reasonably low value. Usually, the SWR reading is higher than normal if the whip antenna faces a metal surface of the car. You can often get a rough check on antenna resonance by moving your hand towards the antenna when the transmitter is on the air and observing the resulting change in the SWR reading. If the reading increases when your hand approaches the antenna, it indicates that the antenna is *too long* and requires shortening to reach resonance. If, on the other hand, the SWR decreases when you bring your hand near the antenna, it indicates that the antenna is *too short* and requires lengthening to reach resonance. The tip of the whip antenna may be lengthened or shortened a bit at a time and this simple test repeated until resonance is established. At this point, the SWR is a minimum value.

Plastic coated helical wound whip antennas (to be discussed later) may be shortened a turn at a time by removing the top plastic cup and unwinding the conductor, cutting off a small amount until the SWR reading is below 2. Go slowly, as it is difficult to restore missing turns!

Any antenna may be lengthened by placing a small coil at the base of the antenna between the bottom of the whip and the center conductor of the coaxial transmission line. The coil can be about 1/2-inch in diameter and can consist of a few turns of insulated, plastic covered house wire. Adjust the number of turns for lowest SWR reading.

Fig. 5 RESONANT FREQUENCY of mobile whip antenna is checked with grip-dip oscillator. One turn coil is placed between whip and car body. Grid-dip oscillator is coupled to coil. When oscillator is tuned to the resonant frequency of the whip, meter of instrument will kick sharply.

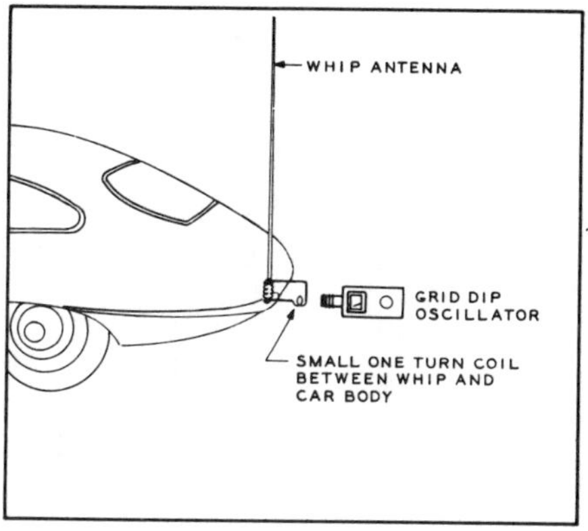

"Grid-Dipping" the Mobile Antenna

Experienced CBers speak knowingly of *grid-dipping* the mobile antenna to achieve superior performance. This means that the resonant frequency of the antenna is measured with the aid of a *grid-dip oscillator* (grid-dip meter). This relatively inexpensive instrument is very helpful for antenna adjustment. Basically, it is a calibrated, low power tunable oscillator equipped with a meter which indicates oscillator activity. When the oscillator is held a few inches from an antenna which absorbs power from the oscillator, the activity of the oscillator decreases and the indicating meter dips, or "kicks", very sharply. This indicates that the grid-dip oscillator is tuned to the resonant frequency of the antenna.

The grid-dip oscillator is coupled to the antenna by means of a one turn coil temporarily placed at the bottom of the antenna, connected between the antenna and the body of the vehicle, as shown in Figure 5. The coaxial transmission line is disconnected at the antenna for this test. The coil of the grid-dip oscillator is placed very close to the one turn coil, or slipped through it, and the dial of the oscillator tuned slowly back and forth across the operating frequency range of the antenna. A sharp dip will be noticed on the meter as the resonant frequency of the antenna is passed. When the dip has been found, the resonant frequency of the antenna may be determined by listening for the grid-dip oscillator signal on a nearby receiver which has an accurate dial calibration. The grid-dip oscillator should be moved away from the one turn coil until a very small dip is observed at resonance. This will insure an accurate reading. If several such readings are made, and the resulting frequency measurements averaged, you can determine the resonant frequency of the whip antenna within a few kHz.

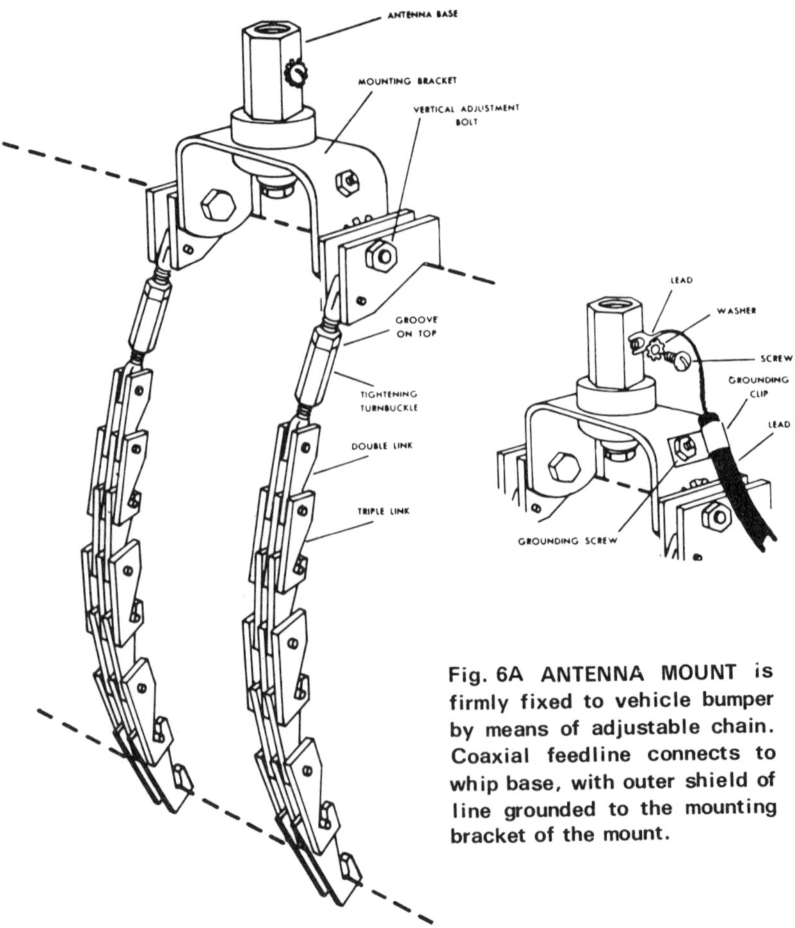

Fig. 6A ANTENNA MOUNT is firmly fixed to vehicle bumper by means of adjustable chain. Coaxial feedline connects to whip base, with outer shield of line grounded to the mounting bracket of the mount.

Detailed operation of a grid-dip meter and additional information on antenna measurements are covered in the operation manual supplied with the instrument.

Antenna Placement On Your Vehicle

Unfortunately, automobiles are not designed with the idea of mounting a CB antenna on them, and often antenna placement depends more upon the sweep of the vehicle body panels than upon electrical efficiency. The unfortunate CBer may reluctantly conclude that the only practical antenna is a roof or rear deck mounted mini-whip, regardless of the lower electrical efficiency of this form of abbreviated antenna.

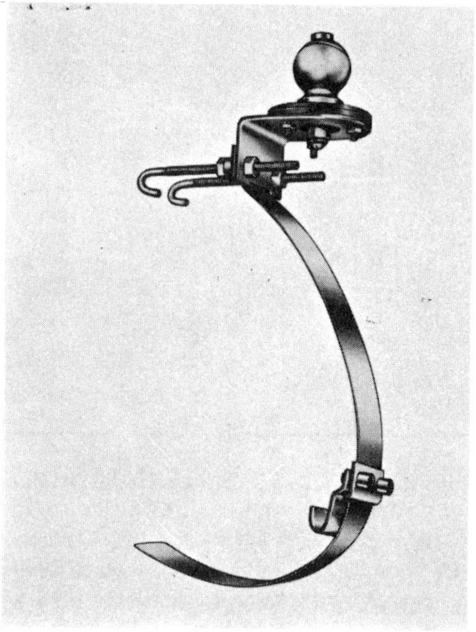

Fig. 6B FLEXIBLE STRAP on bumper mount permits it to be fixed to irregular contours of bumper. The clamps hook over bumper edge without need of drilling holes.

Those operators who place performance above appearance will select the quarter wavelength whip every time!

Quarter wavelength whip antennas, because of their height, are usually mounted low on the vehicle, often on the rear bumper or fender as shown in Figure 6. Chain or strap type mounts are available; they clamp directly over the edges of the bumper without the need of drilling mounting holes in the vehicle. The mount is made up of a flexible strap, or a

Fig. 6C WHIP ANTENNA screws into the socket of this adjustable bumper mount. Care should be taken that whip does not touch trunk lid when car is moving.

TYPICAL MOBILE ANTENNA HARDWARE. At left is a ball and socket mount which allows whip antenna to be mounted on inclined surface of vehicle. Next item is a gutter mount clamp for a temporary antenna installation. At center is an antenna clip to hold whip antenna down to rain gutter when CB gear is not in use. Next item is a mount which attaches to edge of trunk lid. At right is popular, flexible steel spring mount often placed at antenna base to absorb road shocks.

number of chain links that fit around the curvature of the bumper. The antenna is held in a vertical position by an insulated adapter bolted to the top bracket of the mount. Sometimes a heavy spring mounting is used to absorb road shock.

In most cases the spring is not required, although it is useful when driving on rough, "country" roads. In any event, use of the spring won't hurt the overall operation of the CB radio, providing the length of the spring is taken into account when determining overall antenna length.

The Transmission Line

A 50 ohm coaxial transmission line is run from the CB radio installation at the dash to the rear of the vehicle, or wherever the whip is located, and is connected to the mount receptacle, as shown in the drawing. The outer conductor of the line is twisted into a pig-tail and fastened to the mounting bracket (ground), and the center conductor is fastened to the insulated antenna receptacle. Some mounts have a grounding clip for the outer conductor of the line. In this case, the black vinyl jacket of the line must be removed from that portion of the cable that passes under the grounding clip (believe it or not, this is sometimes not done!) If a bolted connection is used, the pig-tail is soldered into the lug placed beneath the bolt. Be certain to make strong,

Fig. 7A DELUXE BALL MOUNT and spring are used to mount whip antenna at an angle on vehicle surface so that whip is vertical regardless of the plane of the mount. Assembly view of mount is shown in Fig. 7B.

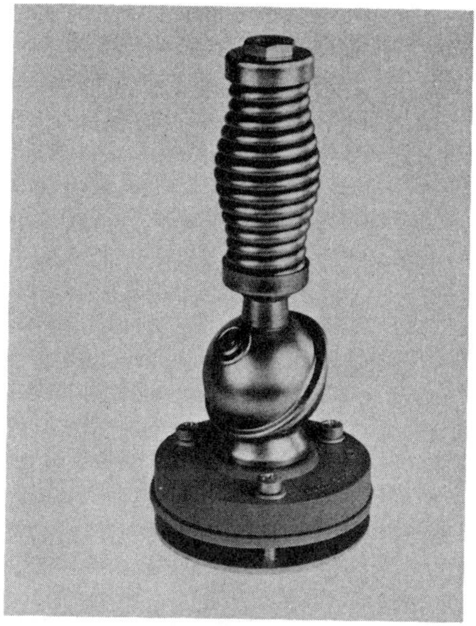

rugged connections at the antenna mount so that they will not come apart under vibration and shock, or provide intermittent connections. Securely tape the exposed end of the coaxial line so that water and road dirt cannot enter it. The coaxial line may be routed into the interior of the vehicle through a small hole drilled in the body at a nearby point. The sharp edge of the hole must be covered with a rubber grommet or plug to prevent the vinyl jacket of the cable from chafing against the body of the car. Running the coaxial line underneath the vehicle for any length is not recommended as it is then exposed to road hazards and dirt.

The whip antenna must remain free and clear of the body of the vehicle. Use of a bumper mount on station wagons, trucks and vans is not recommended as the whip passes too close to the upper metal body panels of the vehicle and severe detuning of the antenna may result. In this situation, a shorter antenna mounted higher on the body or roof of the vehicle is recommended.

The Ball Mount

A *ball mount* and spring (Figure 7) can be used to mount the whip antenna at an angle on the vehicle so that the antenna itself is in a vertical position, regardless of the plane of the mount. Usual placements include the rear deck, the side or top of the fender and (for daring

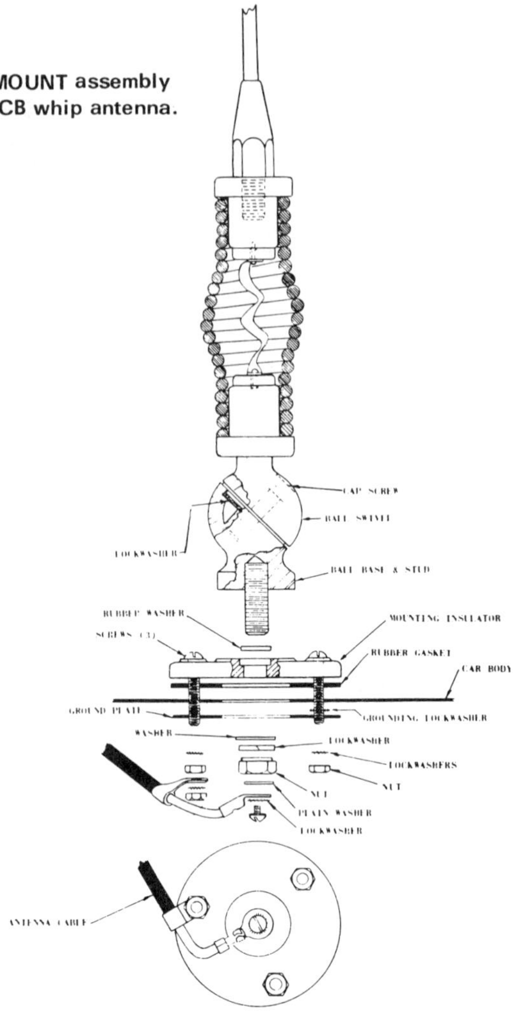

Fig. 7B BALL MOUNT assembly and spring for CB whip antenna.

souls) the top, flat portion of the roof. In the latter case, care must be taken to make sure that the big whip does not strike overhead electrical wires and tree limbs.

The ball mount requires that a mounting hole be drilled in the skin of the vehicle on a relatively flat surface. Installation varies with the type and make of mount, and the procedure outlined by the manufacturer should be followed. Once the mount is in place, the whip is inserted in the socket and the rotary ball joint adjusted to align the whip in a vertical position.

EFFICIENT MINI-WHIP antennas use high-Q loading coil at base. These 54 inch whips and excellent loading coils provide highest possible efficiency with minimum wind resistance. Center loaded whip, theoretically more efficient than base loaded version, often uses low efficiency, compact coil in order to reduce wind resistance.

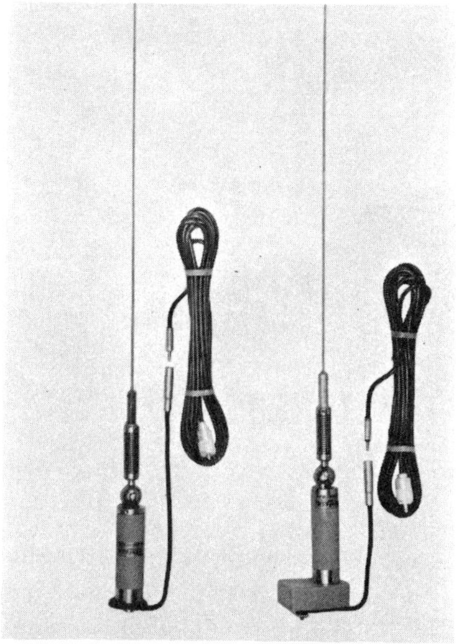

Using A Mini-Whip Antenna

What is the CBer to do who, for aesthetic or practical reasons, cannot mount a full length whip antenna on his vehicle? The only answer is to use a shorter whip antenna that satisfies the imposed limitations. Generally speaking, a center loaded mini-whip will have better efficiency than a base-loaded mini-whip, and a long mini-whip will outperform a short mini-whip. Several manufacturers make a center loaded mini-whip antenna about 54 inches long, and this is a good compromise antenna if the quarter wavelength whip cannot be used. An antenna of this size has a theoretical efficiency of about 50 percent, provided a well-designed loading coil is used. Such a mini-whip has a field pattern that is only -3 decibels below the field pattern of a full size whip antenna. This sacrifice in signal strength is not too drastic and worthwhile performance can be obtained. As the mini-whip grows progressively shorter, and as the loading coil is moved from the center to the base of the antenna, the overall antenna efficiency drops rapidly, as discussed earlier. Extremely short CB base loaded whips (18 inches or so) should only be used as a last resort, as their field pattern is feeble compared to that of a full length antenna. The answer to the antenna problem, then, is to use the longest whip antenna you can safely place on your vehicle, considering the aesthetic restrictions, if any.

Fig. 8A **TRUNK LIP MOUNT** is fastened to vehicle by adjustable fixture which clamps under edge of car body.

Trunk Lip Antenna Mounts

Many CBers hesitate to drill holes in their cars and are interested in an antenna mount that will not scar the body of their automobile. The *trunk lip* mount is a device that meets this need (Figure 8). The adjustable antenna mount is slipped beneath the edge of the trunk lid and bolted firmly to the groove of the car body as shown in the illustration. Enough clearance exists around the edge of most trunk lids to bring a small coaxial line (RG-58A/U) through the gap and up to the antenna mount. The inner conductor of the line is attached firmly to the antenna terminal and the shield is attached solidly to a termination bolt on the lip mounting bracket. Make sure that the mount makes a good electrical connection directly to the metal body of the vehicle--a good connection

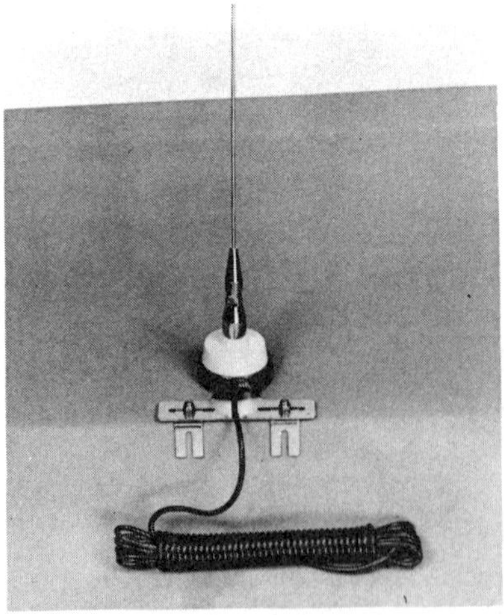

Fig. 8B **TRUNK LIP MOUNT** has small adjustable clamp which fits under edge of lip. It is bolted in place from under the lip so that surface finish of vehicle is undamaged.

IF YOU LOOK CLOSELY in the foreground, you'll notice this petite CB operator is holding a streamlined, trunk lip mount. Girl and mount show great form!

cannot be made through the body paint; scrape off body paint at the point of contact.

Some trunk mounts fasten to the trunk lid, as shown in Figure 9. Set screws grip the lid underneath the mount, which is placed at the rear lip of the lid, near the rear window.

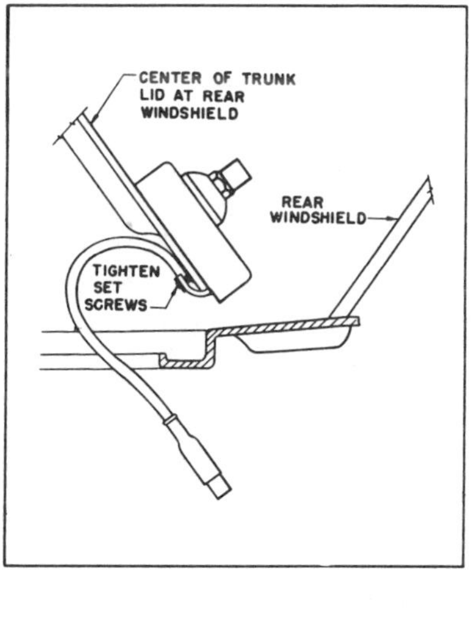

Fig. 9 TRUNK LID MOUNT makes antenna installation easy. Set screw grips lid underneath mount which is placed at edge of lid, near rear window. Paint on underside of lid should be scraped off at point of contact.

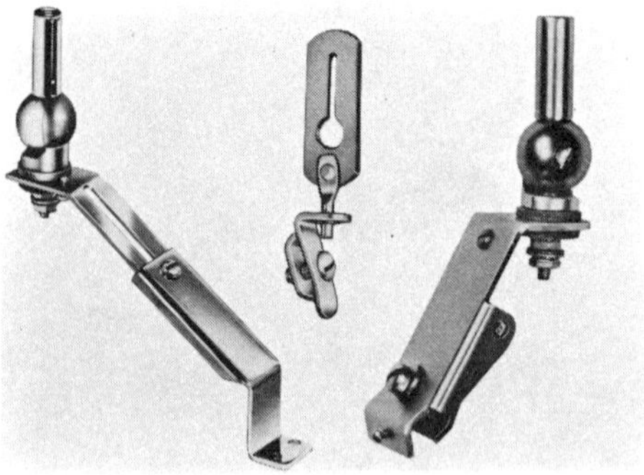

Fig. 10 GUTTER CLAMP MOUNTS. At left is adjustable mount which can be bolted to gutter or edge of trunk lip. Center is gutter clip to hold end of whip when it is bent down for storage. At right is removable gutter clamp mount.

Gutter Clamp Antenna Mounts

A removable mini-whip may be clamped to the rain gutter of the vehicle by means of a *gutter clamp* (Figure 10). The mount is affixed to the outer rim of the rain gutter, taking care to be sure that the retaining screw of the mount breaks through the enamel paint coating of the gutter and makes a good contact to the body of the vehicle. Scraping off the paint at this spot is a good idea. The mount is adjustable to permit placing the antenna in a vertical position.

Adjusting Your Whip Antenna for Best Performance

While the majority of purchased mobile antennas have been pretuned at the factory, the conditions encountered on individual vehicles vary and each antenna installation should be checked for proper performance. This test should be run with the aid of an SWR meter by noting the value of SWR for each channel the antenna is to be used on. Many solid state (transistorized) CB radios require a low value of SWR on the transmission line, otherwise the r-f output transistor may be damaged. CB transmitters using a vacuum tube in the output stage are not so vulnerable in this respect.

The SWR may be checked by placing an SWR meter in the coaxial line near the CB transceiver (see Chapter 10). The SWR should be checked on Channel 1 (or the lowest frequency channel you are using), then checked on each progressively higher channel, up to Channel 40 (or the highest channel you will use). Write down the SWR reading on each channel and then make a plot of SWR readings for the channels, such as the example shown in Figure 4.

Generally speaking, if the SWR reading on Channel 1 (or your lowest channel) is higher than that observed on Channel 23 (or your highest

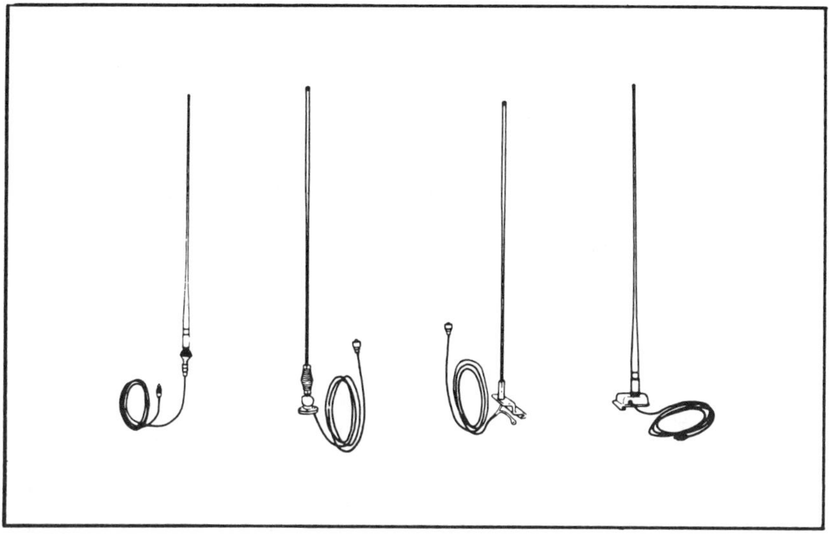

MANY ANTENNAS TO CHOOSE FROM. These are very popular. At left, the cowl mounted antenna, followed by the ball-and-socket mount with spring. Next is the trunk lip antenna and at right the trunk lid (removable) mount.

channel), the antenna is too short. If, on the other hand, the highest frequency channel has a higher value of SWR than the lowest frequency channel, the antenna is too long. With full-size quarter wavelength whips, the variation in SWR across the channels will probably be quite small and--unless the antenna is badly detuned by the body of the vehicle--no adjustment in antenna length is necessary. Mini-whips will usually show a higher value of SWR on all channels than longer antennas, and the shape of the SWR curve will be sharper, as shown in the illustration.

If the shape of the SWR curve is smooth, with minimum value of SWR falling near a mid-frequency channel (say, Channel 20), the antenna may be considered to be properly tuned. Most quarter wavelength whips show an SWR value less than 2 across all CB channels.

An SWR reading of less than 2 indicates the antenna system is probably as good as can be expected without a lot of time spent fiddling with antenna tuning. Such adjustments may lower the SWR a bit, but probably won't make your transmissions any louder at the receiving end of your circuit.

Mini-Whip Adjustments

The mini-whip presents a different situation. The SWR curve of such an antenna is sharper than that of the full length whip and the antenna

DISGUISED WHIP ANTENNA takes place of the standard cowl mounted broadcast antenna. Some antennas are double purpose affairs, permitting broadcast reception as well as CB operation by means of ''signal splitter'' which separates the two signals.

is more adversely affected by the nearby body of the vehicle. Time spent in adjusting the mini-whip will pay big dividends in signal strength reports. Most mini-whips have adjustable tip sections which may be varied to bring the antenna into resonance. Adjustment of antenna length one-quarter inch at a time should help in reducing the SWR value and in producing an SWR curve such as the one illustrated. The minimum value of SWR varies from whip to whip and is also a function of antenna placement on the vehicle. The SWR may run as high as 3 or more on both Channels 1 and 40 when the mini-whip is adjusted for lowest SWR on Channel 19 or 20. The shape of the SWR curve, moreover, is a function of loading coil design and manufacture, and varies widely from one style of whip to the next. In any event, the name of the game is to drop the SWR reading to as low a value as possible on your most-used channel. You probably won't find it possible to get a satisfactory SWR reading on all channels when using a very short mini-whip, regardless of published advertisements. Operation on channels that exhibit a high value of SWR (3 or more) should not be attempted, especially with solid state (transistorized) CB equipment because of possible damage to the r-f output transistor.

If all seems well with your antenna and you still experience an unusually high SWR reading, it may indicate a poor ground connection between the outer conductor (braid) of the transmission line and the vehicle at the base of the antenna, or perhaps detuning of the antenna resulting from capacity between the whip and some vertical part of the car body. As a last resort, it may be necessary to move the antenna to another location on the vehicle in order to achieve a tolerably low SWR value on the transmission line.

With the tremendous number of different automobile makes and body styles available, and the large number of whip antennas manufactured

Fig. 11 PHASED antennas on trucks, recreation vehicles and autos help to reduce signal "flutter" when receiving and improve signals. The irregular body of the vehicle often tends to make transmission in certain directions difficult. Two antennas tend to smooth out signal patterns and provide improved results in all directions. (Antennas by Shakespeare.)

by various companies, it is difficult to provide specific SWR values that would be applicable in all cases. Some whip antennas tend to function better on certain body styles than do others, and the authors know of no reliable information that would aid the prospective buyer in choosing the best whip antenna for his particular vehicle. Suffice it to say that the quarter wavelength whip antenna is the least critical of all available antennas, and it may be used as a "standard of comparison" for other mobile antenna types. Remember--all mini-whip antennas are a compromise and electrical efficiency is sacrificed to a greater or lesser degree in order to shorten the physical length of the antenna. You pays your money and you takes your choice!

Multiple Whip Antennas

Popular with truckers, twin ("co-phased") whip antennas are now being used extensively on automobiles. For best operation, the whips should be spaced at least 5 feet apart--and preferably 8 to 9 feet apart. They are fed with a coaxial cable harness that connects the antennas in parallel. Broadcast stations often use parallel-connected antennas of this type.

Theoretically, two co-phased quarter wavelength whip antennas will double the signal strength in line with the body of the vehicle. This is true if the antennas are spaced correctly, fed properly and placed over an infinite ground screen. The first two criteria are easy to satisfy in mobile service, but the third is almost impossible to achieve. As a result, the power gain of co-phased mobile antennas is usually less than optimum and the directive characteristics common to this style of antenna mounting is diminished.

Even so, the use of two antennas can be advantageous to the mobile operator. The two antennas tend to "iron out" irregularities in the field pattern (see Figure 13) resulting in a more constant communication range that is less dependent upon the position of the vehicle with respect to the receiving station. Fading and signal "flutter" when the vehicle is in motion may also be reduced.

The Helix Whip Antenna

A modified form of loaded antenna is the helically wound whip antenna employing continuous loading (Figure 12). This antenna is composed of a spirally wound coil running the length of the antenna from top to bottom. Some designs have the coil in spaced sections, while others have a continuous winding. Commercially available models are usually made of a tapered, Fibreglas core upon which a narrow copper

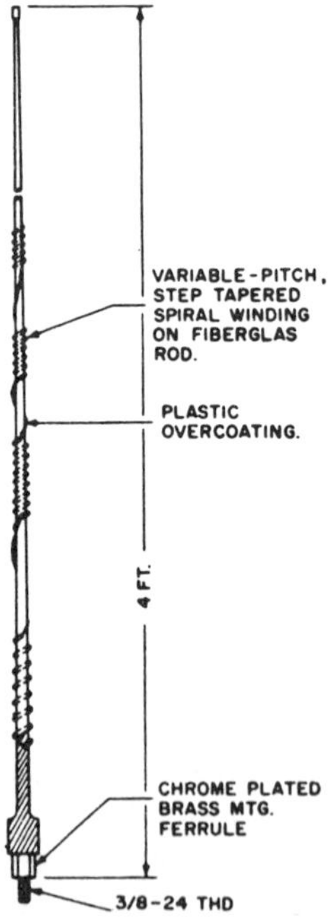

Fig. 12 HELIX WHIP ANTENNA is composed of spirally wound coil running the length of a Fiberglas rod. A variable-pitch winding may be used, or the coil may be wound in sections, as shown here. Winding is covered with waterproof coating of epoxy. Helix whip exhibits relatively high efficiency and is preferred to mini-whip having a low-Q loading coil at the base.

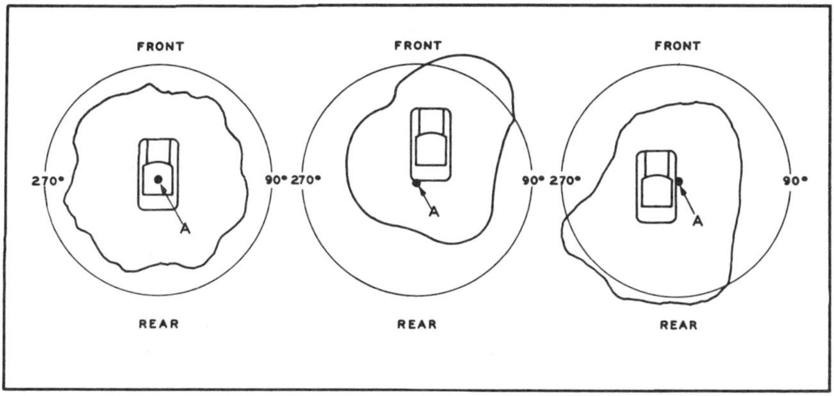

Fig. 13 DIRECTIVITY EFFECT OF MOBILE ANTENNA depends upon mounting position on vehicle. Popular bumper whip (center) is directive towards right of car. Optimum pattern is obtained when antenna is mounted at center of car (left).

ribbon is spirally wound, much in the fashion of a barber's pole. The ribbon is covered with an overcoating of epoxy or other waterproof material for protection against the weather. Continuously loaded whip antennas of this type for the CB channels average about 48 inches long and combine the relatively high efficiency of a full length whip with the compactness of a coil-loaded mini-whip. Their use is recommended where a short CB mobile antenna is a necessity and the losses incurred in a base loaded antenna are not desired.

Whip Antenna Patterns

When a nondirectional whip antenna is mounted on the irregularly-shaped body of a vehicle, the symmetrical antenna pattern is distorted and the whip radiates better in some directions that in others. Figure 13 shows typical radiation patterns for placement of a single whip at various locations on an automobile. In most instances, the vehicle body acts as a sort of director element, with the main lobe of radiation aimed in the direction of the vehicle mass. Thus, a whip mounted on the rear of the vehicle produces the strongest signal toward the front of the vehicle. A whip mounted near the rear, left tail light produces the maximum signal in the direction of the right, front headlight.

Siting Your Car For Greatest Radio Range

Since your CB signals follow the "line of sight path" to a distant station, it stands to reason that you'll get longest range and best signal

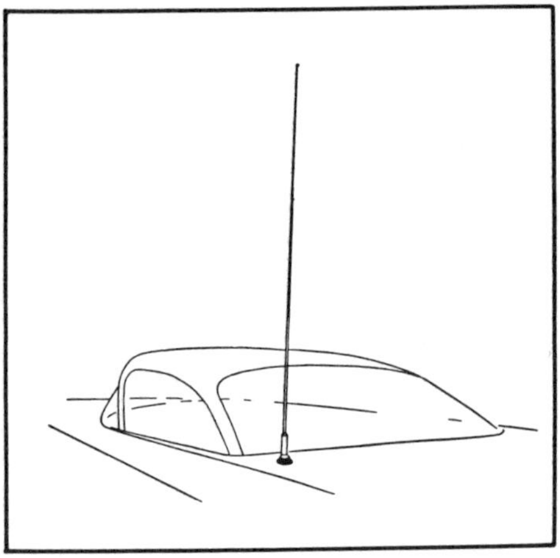

CB MOBILE IS NO BETTER than the antenna. Whip should be protected from overhead objects and base insulator should be cleaned regularly to remove grease and accumulated dirt. Inexpensive SWR meter may be installed in coaxial line to antenna for continuous check during operating periods. Be kind to your antenna and it will work for you!

strength reports when your vehicle is atop a hill as opposed to being in a valley. Many mobile CB operators familiarize themselves with the terrain in their vicinity and make it a point to use their equipment when the vehicle is on high ground, uncluttered with high tension lines and steel buildings. Poor radio results will be obtained when the vehicle is in a valley; surrounded by high buildings; in a tunnel or on a large, metal bridge. Tuned and adjusted for average conditions, the vehicular radio usually performs in a substandard manner under less than average environmental conditions, and the CB operator should not be surprised if, on occasion, he finds that his equipment is "blacked out" because of the surrounding environment.

Chapter 12

CB Antennas for Marine Service

(Sea-B Radio on the Bounding Main)

Multiply the pleasure and safety of your boat by the addition of a CB radio! Thousands of week-end sailors on lakes and seas have installed CB radio for extra enjoyment and safety. Safety, of course, comes first and CB radio can provide you with an inexpensive, important communication link to shore or boat in case of an emergency.

The problem, at the present writing, is that the U.S. Coast Guard does not "guard" (listen to) CB Channel 9, the channel designated by the FCC for CB emergency use. This means that in a marine emergency, your call for assistance would be heard only by other CB operators on shore or perhaps aboard a nearby boat. There is, or course, no guarantee that any CB operator would hear you although many CBers have assisted in emergencies at sea. The Coast Guard *does* continuously guard the HF and VHF marine emergency radio frequencies, and there is a possibility that they may guard the CB emergency frequency at some future date. A sizable pleasure craft which ventures far offshore should not rely only on CB radio for emergencies, but should have aboard marine radiotelephone equipment which can be switched to the regular marine distress frequencies; CB provides a useful adjunct to this equipment, and can be the only channel on which you can talk with a boat in distress equipped only with CB gear.

No boat is too small for CB radio--including a rowboat! Hand held CB transceivers, with powers from 100 milliwatts (a milliwatt is 1/1000 watt) to 5 watts, have a range at sea of a few miles to 10 miles or more if interference is not too heavy, and can be useful in summoning help when equipment for the normal distress frequencies is not aboard.

The six and twelve volt batteries used to start outboard engines can easily power a CB set (if it has the proper power connections) and when you are tied up at dockside, your CB radio will operate from the shore electric lines unless it is entirely operated by dry batteries, or has only a mobile-type power supply. Larger boats, of course, have extensive electrical systems that will power your CB radio with ease.

Installation of CB radio equipment on a boat is usually covered in the general operating instructions for the equipment. Often left unanswered is the problem of installing an efficient and rugged marine CB antenna on the boat. This chapter provides you with suggestions for efficient antenna installations for CB radio on pleasure craft and work boats.

Your Marine Radio Range

CB marine radio is subject to the same Laws of Nature as land-based CB radio. The "Radio Horizon" (see Chapter 3) is usually limited to about 3 to 30 miles at sea, principally because of the low overall height of the antenna which is restricted by boat size and also by F.C.C. regulation to 20 feet above the surface on which the antenna is mounted. This distance is measured from the top of the antenna to the mounting surface. Thus the CB radio range is considerably less than that of the more powerful and expensive high frequency marine radiotelephones and about equal to the VHF marine radio channels. Even so, the CB radio range is great enough so that a marine installation can be a valuable accessory to your boat, and at lower cost than a HF or VHF installation.

CB Marine Antenna Mounting Problems

As with land-based CB radio, the CB marine radio antenna system is the most critical element in the whole installation. It is imperative that the antenna be as efficient as possible, or the tiny output power of your CB set will be wasted. Unlike land-based installations, it is usually impossible to install a beam antenna aboard most pleasure boats and the seagoing CB operator must satisfy himself with relatively simple antennas that will not interfere with the boat's rigging or operation. Sailboats, in particular, carry a maze of rigging which varies from boat to boat. Suspending a vertical CB antenna in the rigging can interfere with the sails and cause antenna detuning and loss of effective signal power unless precautions are taken.

The ideal location for a CB antenna aboard a sailboat is, or course, atop the tallest mast but this is seldom practicable because another antenna may already be there, and because it is a difficult spot to reach.

MARINE RADIO SERVICE. The majority of large pleasure boats use HF marine radio for ship-to-shore communication. A growing trend is to supplement this equipment with CB radio for short distance communication with other boats or to a landbased CB installation. Boat-to-boat CB radio may be used for instant short range communication at sea and the inexpensive CB installation provides invaluable back-up in an emergency. Smaller boats rely upon CB radio exclusively for communication. Luxury yachts shown here have both marine HF radio and CB radio, plus radio direction finder and electronic depth finder.

Some sailboat CB antennas are mounted on a small platform built out to one side of the mast; the antenna is parallel to the mast and 2 to 3 feet from it. It is also possible to use an insulated (sometimes called "isolated") shroud or stay as a CB antenna in conjunction with an *antenna tuner* below decks next to the CB transceiver, but this is somewhat complicated and is not recommended unless there is no alternative. Isolated stay antennas normally are used for the HF marine frequencies and work well when properly tuned.

The power boat, on the other hand, often has no major mast structure on which the CB antenna can be mounted. Both types of vessels, moreover, continually expose the CB antenna system to sea air and moisture that can quickly corrode it if the boat owner is not careful.

CB MARINE ANTENNA may be mounted to side of cabin on pleasure boat. If a quarter wavelength whip antenna is used, two or more radials must be run along hull of boat to provide counterpoise, as described in text. Half-wavelength and five-eighths wavelength vertical antennas require no radials.

Basic Marine Antenna Rules

Regardless of the type or size of boat, or antenna, there are a few basic antenna rules that should be emphasized before you install your marine CB radio equipment.

1- The CB marine antenna should be examined for the possibility of its electrical continuity becoming intermittent or being destroyed by vibration or corrosion. Telescoping antennas should be avoided because of their numerous joints unless they are used only for short periods of time afloat. Antenna joints and connections should, ideally, be bolted together. The use of self-tapping sheet metal screws (which sometimes loosen under vibration) should be avoided.

2- Salt air, salt spray and salt water have a corrosive effect on metals, including those used in antennas, and special precautions should be taken to minimize damage. Your CB marine antenna can be protected against rust and corrosion by painting it, making certain *not* to paint the insulators and antenna base connection. Many paints conduct electricity to an extent, and a painted insulator can easily short out your entire antenna system. Paint on the antenna element itself, however, does no harm. Antennas may be painted with any good quality marine paint. *Rustoleum* paint, in particular, does a good job of protecting your antenna.

After a particularly wet time afloat in salt water, it is a good idea to wash down all antennas with a fresh water hose (and the boat, too, as experienced yachtsmen know).

The use of electronic equipment on fresh water lakes and rivers poses fewer problems since there is no salt to cause troubles. Here, your principal interest is in making, and maintaining, solid watertight connections at all points, including feedline to antenna base connector.

3- A particular source of potential trouble in marine antenna installations lies in fastening the transmission line, or lead-in, to the antenna. First, be sure to scrape bright with a knife the lug or wire to be fastened to the antenna base, also that portion of the base which makes contact with the lug or wire. This removes oxidation and corrosion which can result in a poor, intermittent electrical contact. Second, fasten the transmission line to the antenna tightly with bolt, lock washer and nut. Third, waterproof this connection either with 3 to 4 layers of tightly wound plastic electrical tape, or better yet, with the special waterproofing compound made by the large chemical companies, or with General Electric's *RTV-102* silicone sealant which comes in small tubes. (Note: these silicone sealants are difficult to remove and should be used primarily for permanent or semi-permanent installations).

4- The CB marine antenna should be mounted as high as possible on the boat and as far away from large metallic surfaces, rigging and electrical wires as possible. Not only will this provide the greatest radio range, but it prevents detuning and also means that the antenna receives less salt spray and green water.

5- CB marine antenna construction should be as light and sturdy as possible. A heavy antenna may break under continuous pounding and go over the side as a result of sea motion -- often just when it is needed most!

The mounting base should be secured to the boat with through-bolts rather than with wood screws, which may pull out under stress. Mounting hardware installed in areas exposed to sea spray should be coated with plastic or weather-resistant compound to prevent water from seeping under the mount and causing rot or corrosion.

6- Do not attach flags, pennants, or clothing (to dry!) to your CB antenna--or to any other antenna; these objects often become soaked with spray and water and seriously detune the antenna, with the result that your radio works poorly, if at all. (One powerboat skipper whose 100-watt HF marine radiotelephone would not operate at all one wet afternoon on Nantucket Sound found that his trouble was a Race Committee pennant fastened to his antenna which was drenched with sea water all afternoon).

7- Do not fasten wires or lines to your CB antenna (or any antenna) since they,too,detune the antenna and adversely affect its performance.

8- Antennas which must be lowered to pass under a bridge should be hinged and have a quick release "pelican hook" spliced into the forward guy. Guy wires, if used, must be insulated from the antenna itself by small ceramic insulators. Self-supporting whip antennas, moreover, require a ball-and-socket mount for lowering purposes.

Electrolytic and Galvanic Corrosion

When you enjoy the pleasures of boating, you encounter two phenomena which may be new to you: (1) *Electrolytic corrosion*, and (2) *Galvanic corrosion*. While these types of corrosion are usually not related directly to CB radio operation afloat, it is a good idea to know something about them since they are the subject of great confusion to even some experienced yachtsmen.

1- Electrolytic corrosion is the deterioration of metals immersed in salt water which results from an external electric current (usually from the boat's battery, electric system, shore power connection, or radio system), coming into contact with metal surfaces.

2- Galvanic corrosion results when two metals of different composition immersed in salt water generate a small electric current between them, even though no external current is applied to either metal.

Both electrolytic corrosion and galvanic corrosion eat away underwater metallic surfaces fairly rapidly. Small metallic particles actually become detached from the metal and pass into the sea water with the result that the metal becomes pockmarked, gets holes in it, and ultimately wears away.

Some metals deteriorate faster than others when subjected to corrosive currents, or even to salt laden air. The best marine fittings are made of stainless steel, monel or bronze, which keeps corrosion at a minimum. Marine radio antennas are sometimes made of stainless steel and are often protected against corrosion by a plastic coating or marine paint.

RUGGED MARINE ANTENNA is built to withstand salt spray, vibration, and rough weather. Base of antenna is securely fastened to deck or other surface by means of metal flange. A swivel joint above the flange allows antenna to be lowered to the deck to pass under bridges or other obstacles. The whip antenna is fastened to superstructure by a special U-clamp which may be removed quickly. The antenna itself is made in two sections of heavy-wall tapered tubing which is given a weatherproofing coat of expoy to protect the metal from the salt spray. Antenna tip is covered with a plastic ball to reduce corona discharge. CB marine antennas are built in this fashion for long life, and to make sure the antenna works when you want it to!

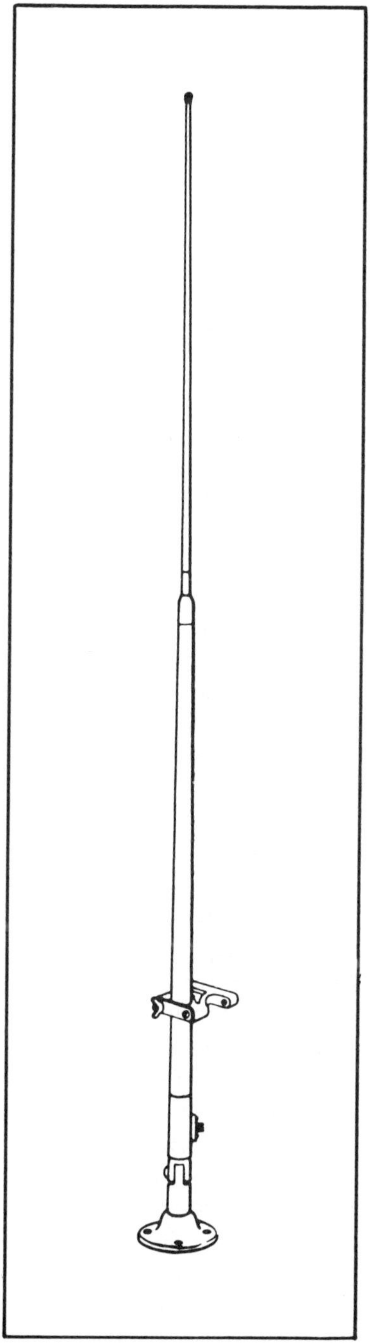

Here is a list of metals in order of *increasing* susceptibility to galvanic corrosion:

1. Stainless steel
2. Monel
3. Bronze
4. Copper
5. Brass
6. Lead
7. Steel
8. Iron
9. Aluminum
10. Galvanized steel
11. Galvanized iron
12. Zinc

In other words, stainless steel resists corrosion the best, zinc the worst.

How to Take Protective Steps

As mentioned, marine antennas can be protected with plastic coating or marine paint, taking care not to coat the insulators or the base connection (which should be thoroughly water-proofed, as described).

The following serve to retard underwater corrosion:

1- Paint all underwater metal fittings and surfaces with high quality anti-fouling marine paint (which often costs more than bonded bourbon!). It may be necessary to scrape, sand and repaint several times a season depending upon the water in which the boat operates.

2- All through-hull fittings, chain plates, rigging, metal masts, engine, and underwater ground plate should be connected together with heavy copper wire (#8 or larger). This places all these metal parts at the same electrical potential and greatly reduces or eliminates corrosive action. (Note: boats with all-metal hulls, such as aluminum or steel, present special problems which are beyond the scope of this chapter).

3- Sometimes special zinc underwater fittings are installed with the knowledge that they will corrode away much faster than fittings of other metals (true), can be replaced when they are badly corroded (true), and that this protects the other underwater metal parts (true). A zinc collar on a propeller shaft is an example which you will see quite often; certain outboard motors have small zinc parts with this theory in mind.

It may be helpful to mention two other related points which, unfortunately, arise more often than they should:

1- When a boat docks at a marina and its electrical system is plugged into shore power, it is important to get the connections (sometimes termed "polarity") right, i.e., be certain that the boat's ground connection is the same as the ground connection in the shore power line plug. *Failure to do this may result in severe electrical shock and possible injury or even death.*

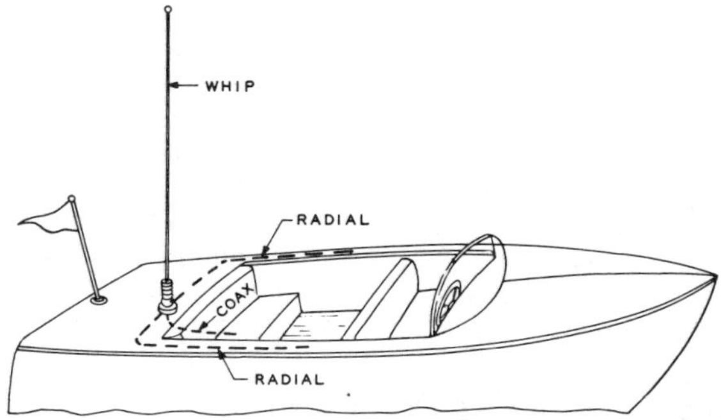

Fig. 1 COUNTERPOISE GROUND may be installed on boat for proper operation of quarter wavelength CB whip antenna. Two or more insulated wires are connected to the outer shielded braid of the coaxial feed line at the base of the antenna. Wires are run along deck of the boat, or strung along the superstructure. Counterpoise wires should lie in horizontal plane and are moved about experimentally until lowest SWR is observed on coaxial line to CB equipment.

2- In a marina, if you are unlucky enough to be tied up next to a boat with a fouled-up electrical system -- and there are more around than you might think! -- it is possible for his mixed-up system to give you some surprises aboard your vessel. The solution? Help him solve his problem!

The Counterpoise Ground

As discussed earlier in this Handbook extended 27 MHz CB whip antennas, such as the half-wavelength or five-eighths wavelength antenna, require no radial or ground connection for operation. Such antennas, when used for marine CB service, are very efficient. Quarter-wavelength whips, on the other hand, require some form of radial system for proper operation. A simple *counterpoise ground,* or simple radial system, may be made out of two or more insulated wires a quarter-wavelength long (about 108 inches at 27 MHz) placed at the base of the whip antenna (Figure 1). The wires are strung out horizontally over the superstructure of the boat. If the hull of the boat is wood or Fiberglas, the

SAILBOAT ANTENNA is placed atop main mast on small bracket. If CB antenna is mounted in this position, radial wires may be run down the mast among the rigging. High whip antenna also serves as lightning rod to protect the boat during storms.

insulated radial wires may be temporarily taped to the boat's deck or superstructure, or laced to convenient points with twine. Care must be taken to keep the radials out of the way of the passengers and crew so that no one will trip over a radial wire and perhaps injure themselves or fall overboard. At the antenna base, the radial wires are connected to the outer braid of the coaxial feedline of the CB installation.

Placement of the radial wires is not critical, and they may be suspended in air or taped to the deck in an out-of-the-way location. The wires need not run in a straight line, nor need they run in an exact horizontal plane. Moving the wires about while watching the SWR meter in the transmission line to the CB transmitter will quickly show optimum placement of the radials that make up the counterpoise ground. Here, as with every type of antenna, your objective is to achieve the lowest possible reading of reflected RF power from the antenna back into your transmitter. When reflected power is at a minimum (as close to 1, or unity, as you can get it), maximum power is transferred from the transmitter to the antenna. Moving the radials about may lower the SWR reading on the meter, and this should be tried when the installation is first made.

A Marine Ground Connection

Most marine radio installations, particularly the HF radio systems, use a *marine ground* which consists of a large copper ground plate

CB AND MARINE radio antennas, plus radar dish are mounted on this luxury power boat. Antennas are mounted to side walls of bridge by rugged brass fittings.

fastened below the waterline of the boat. While essential for HF radio (2 to 16 MHz), the ground installation is rarely needed either for CB radio or VHF radio installations. When used, the marine ground consists of copper sheeting or screen secured to the outside of the hull underneath the waterline with bronze screws. A heavy, insulated wire or strap runs from the ground plate to the ground connection of the high frequency radio. Commercial marine grounds are available in the form of plates, screens and metal tubes, some of which are of doubtful efficiency. Inboard boats sometimes use the engine and drive system as a marine ground (which is not ideal). The CB marine radio may be attached to the marine ground, if desired, by a short, heavy lead (#8 wire, for example) but use of such a ground is not essential and rarely helps the operation of the CB radio afloat because the ground lead is usually long in terms of the CB radio wavelength. To be effective, the lead from the CB radio to the marine ground plate should be very short -- less than a foot or two in length.

What Antenna To Use?

Because of the infinite variety of rigging on pleasure craft, the choice of CB antenna must be left up to the individual boat owner. Generally speaking, the best radio range will be obtained either with a half-wavelength or five-eighths wavelength whip antenna mounted high on a mast and as clear of the boat's rigging as possible. A quarter-wavelength whip, mounted atop the mast, with two or three insulated radials

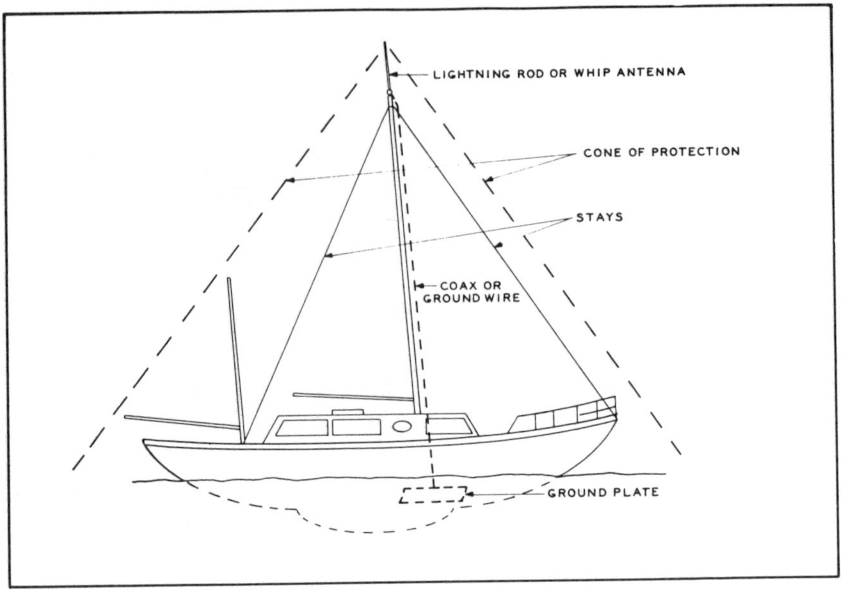

Fig. 2 CB MARINE ANTENNA serves as a lightning rod if lightning arrestor is used and outer conductor of line connected to marine ground with a short, heavy strap. A "cone of protection" about the antenna lessens the chance of a direct hit by lightning. Antenna should be disconnected from CB equipment during storm to protect gear from static discharges.

drooping down into the rigging makes an excellent antenna. A short mini-whip antenna mounted on the hull of the boat will provide the shortest radio range because of its limited height and low electrical efficiency. However, for operation in intracoastal waterways, lakes and close to shore, it does a good job over short ranges.

Lightning Protection

Regardless of antenna choice, the boat (like your house) must be protected from an accidental hit by lightning or a lightning discharge. A grounded point at the top of your highest mast theoretically gives a cone of protection about the mast which includes most of your boat (Figure 2), thus lessening the chance of a direct hit by a lightning bolt. Some boat owners install a lightning rod or pointed wire about one foot above the top of the highest mast with a heavy wire conductor running from the rod to the underwater ground plate on the hull of the boat.

Your CB marine antenna, if the highest point on the boat, may serve as a lightning rod if the braided outer conductor is connected to the marine ground with a short, heavy conductor. Incorporation of a lightning arrestor, such as described in Chapter 16, is a good safety precaution.

Common sense indicates that the CB marine antenna should be disconnected from the equipment to prevent damaging it during a lightning storm, and that those aboard should stand clear of antenna and lead-in during the storm.

CB Radio Afloat

There's a definite place for CB communications aboard boats of all types -- from rowboats to stately yachts -- to add to pleasure and safety afloat. However, the problems of installation and maintenance are different from land-based stations and the special precautions covered in this chapter should be taken to insure reliable communications.

To repeat: larger craft should not rely only on CB radio for safety; they should also have installed regular HF or VHF equipment which covers the distress channels monitored day and night by the U.S. Coast Guard and by vessels at sea. Help can be summoned more quickly and reliably using one of the regular safety frequencies; however, in the case of smaller craft without more elaborate radio equipment, CB radio could spell the difference between safety and tragedy at sea, and on inland rivers and lakes. CB radio is also valuable back-up equipment in case the regular marine radio equipment fails to function at a critical time. And, as so often stressed in this Handbook, the CB marine radio antenna is the key to better and more reliable communications, just as its land-based counterpart.

MARINE RADIO ANTENNA is mounted to superstructure of this power boat. Base of the antenna is seated on a ceramic deck insulator, with coaxial feedline passing through cabin wall. Whip is braced at roofline with side mounted insulator having quick disconnect feature to permit antenna to be lowered when passing under bridges.

"What do you mean I don't understand the Truth Table?"

Chapter 13

CB Antennas You Can Build

(Construction Projects for the CBer)

The vertical antenna is the most popular installation for the majority of CBers and, when properly installed, performs in a very satisfactory manner. Most CB vertical antennas are of the ground plane variety having three or four horizontal radials and are fed with inexpensive coaxial line (usually RG-58A/U).

Where To Place Your Antenna

The CB vertical antenna should be placed as high and as in the clear as possible (consistent with the antenna height limitations imposed by the FCC). The antenna should be mounted away from metallic objects such as drain pipes, rain gutters, telephone wires and utility power lines. These conducting objects may produce severe alterations in the electrical characteristics of the antenna, distorting the pattern and raising the SWR on the transmission line. In any event, power lines can be dangerous and it is wise to stay well clear of them. Metallic objects at moderate distances from the antenna will not necessarily affect operation; in city areas trial and error placement of the antenna may be necessary to avoid "dead spots" in reception and transmission caused by blockage of the radio waves by a nearby obstruction. A little time spent in the choice of antenna site may pay big dividends in better communication.

Bracing wires used to support and steady an antenna are called *guy wires*. Galvanized iron wire is often employed for guys as it is strong and inexpensive, although in time it rusts and must be replaced. The

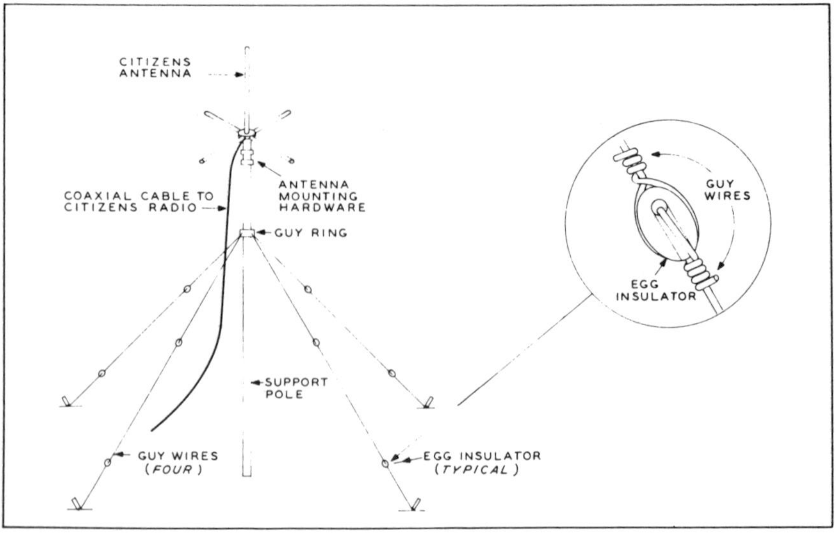

Fig. 1 **STURDY CB INSTALLATION requires three or four guy wires to support the antenna structure. Guy wires are broken into short sections by egg insulators, as shown. Ordinary TV antenna hardware (masts, guy rings, insulators, etc.) may be used for inexpensive CB antennas. A ship-shape installation keeps your antenna in the air in spite of buffeting by winds.**

newer aluminum clothesline wire is satisfactory for guys, is noncorrosive and easier to use than steel wire, and is recommended for CB antennas. To prevent guy wires from acting as unwanted antennas, it is necessary to break the guys into short, isolated sections. Special ceramic insulators *(egg insulators)* are made and used for this purpose. For CB antenna systems, guy wires are usually cut into 6 foot lengths and are spliced together with egg insulators as shown in Figure 1.

Build Your Ground Plane Antenna

Ground plane antenna kits are available at prices from $10 up. You can, however, build your own ground plane antenna for less than five dollars out of readily obtainable parts, if you wish. Here are the plans for a simple antenna that will do a good job for you. A sketch of the complete installation is shown in Figure 2. The assembly consists of a quarter-wavelength vertical aluminum pipe antenna mounted atop four quarter-wavelength semi-horizontal wire radials which also serve as guying supports for a short wooden mast.

The vertical antenna section is made of a length of 3/4-inch dia-

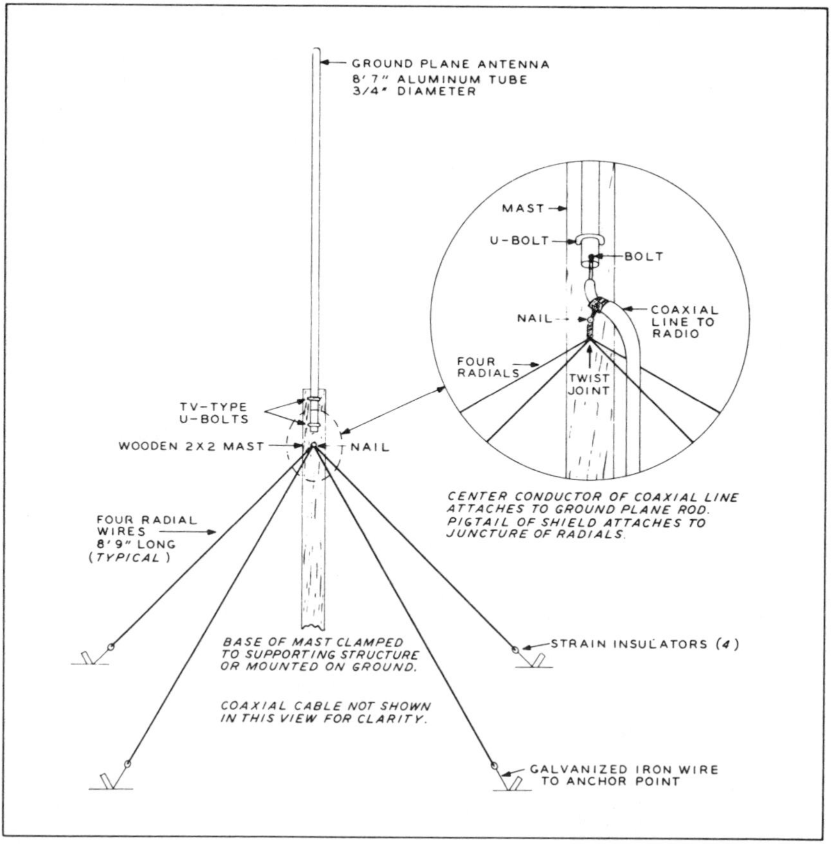

Fig. 2 INEXPENSIVE GROUND PLANE ANTENNA is made of a piece of aluminum tubing and four wire radials. Radials serve as guy wires for the supporting mast. Carefully tape end of line to prevent moisture from entering it. Aluminum clothesline wire extensions may be used at ends of guy wires.

meter aluminum tubing cut to a length of 8'7''. Tubing diameter is not critical. Two small holes are drilled through the tube near one end so that it may be fastened to a wooden mast or support with wood screws. The mast is given two coats of outdoor paint to protect it from moisture before the assembly begins.

Four 9'6'' copper wires are cut and the radials made up next. Clean the insulation from the ends of the wires for a distance of six inches. One end of each wire is passed three inches through the eye of an egg insulator and then looped back and twisted upon itself. The joint is then soldered. At the mast, the free ends of the four radials are twisted together for about three inches, soldered to form a single, heavy lead, and firmly twisted about a heavy nail driven into the wooden mast an inch or so below the lower end of the vertical aluminum tube. Overall radial length should be 8'9'' from nail to egg insulator. A short length of copper wire is looped about the mast and the ends are wrapped around the joint of the radials to serve as a safety wire if the nail should accidentally pull free of the wood. Wires or ropes may be tied to the free "eye" of each egg insulator at the tips of the radials, making them serve as supporting guy wires for the mast.

Preparing the Feedline

Once the vertical antenna section and the radials have been fastened to the mast, the RG-58A/U coaxial feedline should be prepared and soldered in place. If care is taken, the line may be attached directly to the vertical antenna and radials without the need of a coaxial fitting. Here is how you do this:

The outer insulation is removed from the end of the coaxial line with a sharp knife, as follows. The insulation is carefully slit in a circle about 3 inches from the end of the line. Next, the jacket is slit from this circular cut toward the end of the cable, as described in Chapter 9. The slug of vinyl insulation is removed, exposing the outer, braided conductor of the line. The braid is pushed back upon itself to loosen the weave. A sharp tool, such as a nail or awl, is used to open up a hole in the braid about a half-inch from the insulating jacket. Separate the fine wires with care, using the point of the tool until a hole nearly equal in diameter to the diameter of the cable is made (Figure 3). Now, the cable is bent sharply at the hole and the insulated inner conductor is "fished" through the hole you have just made, using the nail or awl as a sort of hook. Once the inner conductor has been carefully pulled through the hole in the braid, the empty braid may be flattened into a pig-tail lead, ready for connection to the antenna radials. All of this fiddling around takes longer to read than to do in practice, and the photograph is worth a thousand words.

The last step is to strip the insulation from the end of the center conductor, tin the conductor and solder it to a husky lug. The lug is bolted to the bottom end of the aluminum tube, making sure the surface of the tube is cleaned of paint and scale at the point of connection. The pig-tail lead of the coaxial line is soldered to the common joint of the four radial wires. The exposed end of the coaxial line, center conductor and pig-tail are now firmly wrapped with vinyl electrical tape, which is given a coat of moisture resistant sealant, such as *General Electric RTV-102*. This material is a rubbery gook, resembling bath-tub

Fig. 3 A PIG-TAIL LEAD MADE THE EASY WAY. Left, the outer jacket is removed from the cable. Next, a hole is picked in the outer braid just above the jacket. The inner conductor is carefully fished through the hole (center). Right, the completed pig-tail lead ready to use.

calk, which does a good job of sealing antenna connections. You can buy it in a plastic tube at most hardware stores. (Common bath-tub calk does a good job, too). Whichever sealant you use, apply it liberally over the joint to make sure that water cannot penetrate and run down the inside of the coaxial line (which happens more often that you would think possible). Water soaked coaxial line makes for very weak signals!

Erecting the Ground Plane Antenna

Place the antenna in a vertical position and fan out the radial/guy wires below it. Adjust the height of the wooden mast so that the wires drop down at about a 45 degree angle (not critical) from a horizontal plane. The wire or rope through the egg insulator on each radial is now tied to a convenient support and the base of the wooden mast is lashed down so that it will not move about. The transmission line is dropped down the pole until it clears the bottom of the radials after which it may be led away in a horizontal direction. The antenna is now ready for use.

The Sloping Radials

The sloping radials serve two purposes in this antenna. First, they act as guy wires. Second, they act as a simple matching transformer which provides a near-perfect match between the ground plane antenna and the 50 ohm coaxial line. The radiation resistance of a true ground plane antenna (one having horizontal radials) is about 30 ohms. The lowest value of SWR that may be achieved in this case is the ratio of line impedance to the antenna radiation resistance. This ratio is 50/30, or an SWR of 1.66. By drooping the radials, the radiation resistance of the antenna is raised, and when they are at an angle of about 45 degrees, the radiation resistance of the modified ground plane will be very close to 50 ohms, accurately matching the characteristics of the transmission line and reducing the SWR to less than 1.2 or so. Thus, maximum power is transferred from transmitter to antenna, and radiated into space.

Build a Cobra Vertical CB Antenna

Are you interested in a simple, inexpensive vertical dipole antenna that can be built in an hour or so, and provides good performance on all CB channels? Useful for emergencies or portable work, the Cobra antenna was named by an enthusiastic user who saw in the r-f choke coil and vertical section a resemblance to the weaving reptile of the snake charmer! What an imagination!

The Cobra antenna is a vertical half-wave dipole composed of an upper quarter-wavelength section made of copper wire and a lower quarter-wavelength section made up of the braided outer conductor of the coaxial line (Figure 4). The simple antenna is suspended at the top from an insulator and a length of rope and hung from a tree or other handy support. The Cobra is fed at the base by a coaxial line, the end of which serves as the bottom portion of the antenna. The remainder of the transmission line is isolated from the antenna portion by a home made choke coil made of a length of the line wound around an inexpensive ferrite core. The Cobra dipole antenna provides a useful power gain of about 1.8 decibel over the ground plane, as shown in the Truth Table.

Antenna Assembly

Your first job is to construct the lower antenna section out of a length of RG-58A/U coaxial line. A PL-259 style plug and UG-175/U

Fig. 4 THE COBRA ANTENNA is ideal temporary or portable installation. Supported at the top, the Cobra may be hoisted into a tree or hung from handy structure. Ferrite coil isolates antenna from the coaxial feedline. Core is Indiana General CF-111.

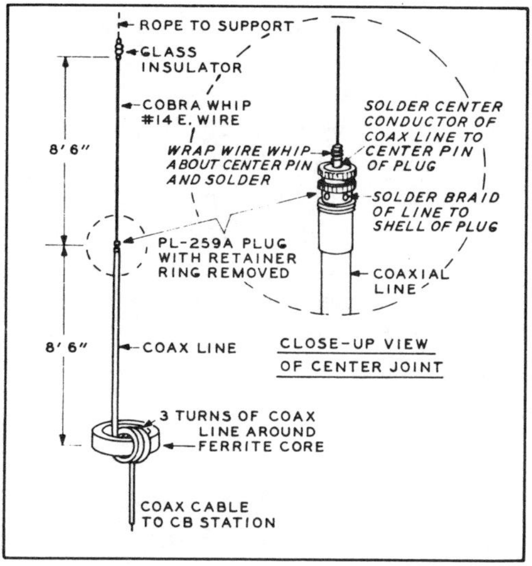

reduction adapter are placed on the end of the line, but the threaded retainer ring of the plug is not used. The plug merely serves as a convenient terminating device for the end of the cable. Trim the line carefully, as explained earlier in this Handbook, and solder the outer braid of the cable to the shell of the plug through the four small solder holes in the plug. Use a very hot iron or gun with a small tip for this job. Solder the center wire of the cable to the center pin of the plug (after you review Chapter 9). Next, the ferrite core is placed on the line exactly 8'6'' down from the plug. A three turn coil is made about the core by passing the line through the core three times. The coil should be about four inches in diameter. Tape the turns together and firmly to the core, which is quite fragile. Now, cut a length of copper wire about ten feet long for the upper section of the antenna. Clean and solder one end of the wire to the center pin of the coaxial plug, wrapping the wire several times around the pin to make a good mechanical joint. Attach the top insulator to the wire so that the distance between the tip of the wire and coaxial plug is 8'6''. Wrap the wire securely back upon itself. The last step is to wrap the coaxial plug center joint with vinyl tape and coat it with RTV waterproofing compound.

Antenna Installation

When you are finished the Cobra is ready to work. It can be hoisted into a nearby tree with the aid of a string and a stone. Tie a light string to the stone and (watching out for nearby windows or spectators) toss the stone over the highest branches of the tree. A husky CBer can usually make a sixty foot toss with ease. (How muscular are you?) Once the string is safely over the branches and the rock is down again to ground level, a heavier rope may be pulled over the branches. The top insulator of the Cobra antenna is attached to the rope and the antenna hoisted up as far as it will go, swinging in a vertical position like a gigantic icicle on a Christmas Tree! The Cobra operates well over all CB channels with an SWR reading of between 1.3 and 1.8 and is a handy, compact emergency antenna to carry in your automobile at all times.

Build a Turnstile Antenna for CB

The horizontal dipole antenna discussed earlier in this Handbook has a "figure-8" pattern of radiation, with the lobes broadside to the line of the wire. Radiation from the ends of the dipole is negligible. Two horizontal dipoles placed at right angles to each other provide a nondirectional pattern, the radiation field of one dipole filling in the nulls of the other. Two properly placed dipoles interconnected by a transmission line are called a *Turnstile* antenna. The turnstile antenna will provide omnidirectional, horizontally polarized radiation, which may be of great benefit to CBers using horizontal polarization to escape interference from nearby vertically polarized operators.

Antenna Assembly

The simplest assembly for a Turnstile antenna uses aluminum tubing as the dipole elements. The tubing is self-supporting and the whole antenna may be attached to a single mast placed at the center of the dipoles. If a metal mast is used, care should be taken that the mast does not make electrical connection with either of the dipoles.

It is important that the dipole sections of the Turnstile antenna receive radio energy in the proper sequence. It is therefore necessary to employ a length of parallel wire "TV-type" ribbon line as an interconnecting section between the dipoles, and also a short length of 50 ohm coaxial line placed between the dipoles and the 70 ohm coaxial line to the station. The interconnecting line and the 50 ohm coaxial line section form electrical matching transformers which distribute the radio energy in the proper fashion. The ribbon line is designated as *Segment A* in Figure 5. One dipole is connected to the station transmission line through a length of 50 ohm (RG-58A/U) coaxial cable *(Segment B)* which

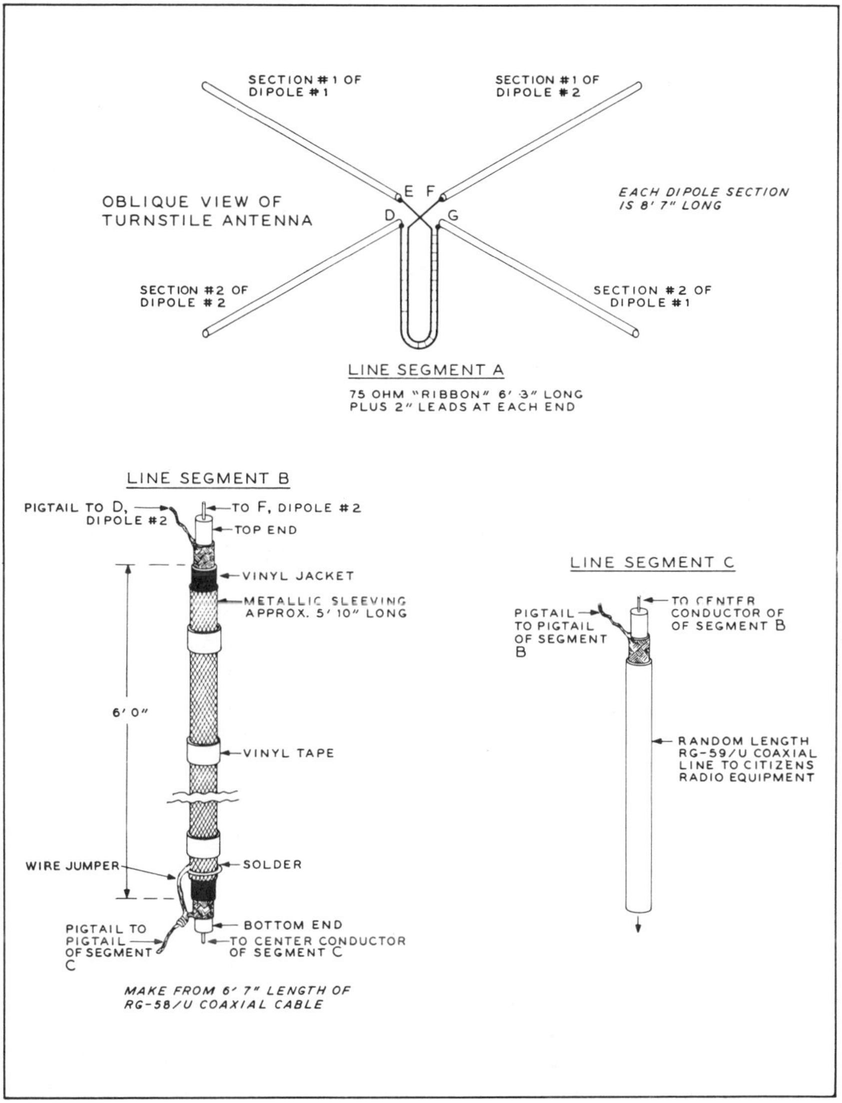

Fig. 5 TURNSTILE ANTENNA provides omnidirectional horizontal polarization for CB service. The antenna consists of two dipoles mounted at right angles (top view). The four dipole sections are fastened to a foot square plywood plate by means of U-bolts. The dipole sections and bolts should not touch the supporting structure, or mast, if it is made of metal. Line segments A and B drop down beneath the dipole sections. Segment A is made of 75 ohm TV line (Amphenol 214-023) and spacing between folded sections is 6 inches.

serves as a transformer and balancing sleeve. This, in turn, connects to the random length 70 ohm transmission line (RG-59A/U) designated as *Segment C* which runs to the CB equipment.The use of proper lengths for line segments A and B, as given, insures that antenna operation is correct.

Use of a 70 ohm transmission line might seem to be heresy, but the great majority of CB equipments will function perfectly with a 70 ohm line in place of the more commonly used 50 ohm line. Using the 70 ohm line solves a rather complex matching problem and allows the antenna to be built and put into use without critical adjustments or the use of an antenna tuner. (Note: the common SWR meter used for 50 ohm line will not give correct SWR readings when used with a 70 ohm line).

Making the Line Segments

Line segment A is made as follows: The two conductor ribbon line is cut to a length of 6'5'' and two inches of insulation are carefully removed from each end of the line, exposing the two copper wires. The end of each wire lead is twisted and tinned to prevent the individual strands from unraveling. This line segment is now connected between the inner tips of the dipoles as follows: One end of the line is attached to the inner mounting bolts (D and F) of dipole #2. The opposite end of the line is attached to the inner mounting bolts (E and G) of dipole #1. Observe in the drawing that the ribbon line should not be twisted or transposed or the wrong connections will be made to dipole #2. Make sure that section #1 of dipole #1 is connected to section #1 of dipole #2 by tracing along the wires of the ribbon line. When the connections have been made the line segment may hang loosely beneath the Turnstile antenna.

Line segment B is made of a section of RG-58A/U or RG-58/U (they are equivalent) coaxial line 6'6'' long, prepared in the manner described in the previous section on the Ground Plane antenna. Two inch leads are made at each end of the line, one lead being a pig-tail formed from the braid, and the other lead being the center conductor of the line.

The Balancing Sleeve

A balancing sleeve is passed over line segment B. This sleeve is made of a 5'10'' length of flexible, braided metallic sleeving which can be purchased at an electronics supply store. The top end of the sleeve reaches to within an inch or two of the Turnstile antenna but makes no electrical connection to it. The bottom end of the sleeve makes electrical contact to the braid of the line segment.

Smooth out the braid and pass it over the coaxial line, bunching it slightly to enlarge its diameter to guide it easily over the end of the line. Once the entire length of braid is on the cable you can easily slip it up to the antenna end of the segment. Trim the top end of the braid with scissors to make it smooth and even and tape it securely to the the outer vinyl jacket of the coaxial line segment so that the end of the braid is about one inch below the termination of the pig-tail. Smooth the braid down along the line until it makes a close fit, wrapping it every six inches or so with a few turns of vinyl tape. The bottom end of the braid should be trimmed even with the end of the line segment.

The next step is to take a three inch piece of scrap hookup wire and strip the insulation from it. Wrap one end around the bottom of the balancing sleeve, folding a fraction of an inch of the braid back over the wire. Solder this connection. Take care not to melt the plastic jacket of the coaxial line segment. You now attach the end of this lead to the pigtail from the outer shield of the line segment, as shown in the illustration. Solder this joint. Segment B is now completed and the top end is attached to mounting bolts D and F of dipole #2, in parallel with the connections of segment A. All joints should now be wrapped with vinyl tape and given a coat of insulating gook (RTV-102 or equivalent) to waterproof them.

The last step is to prepare the transmission line, shown as segment C in the illustration. One end of the 70 ohm line is prepared with a pigtail as previously described and the conductors are connected to the corresponding conductors of segment B, as shown. The connections are soldered, taped and waterproofed. Make sure the connections do not short out the inner conductor to the shield. Wrap each splice separately, then over-wrap both splices with a second length of vinyl tape.

Erecting the Turnstile Antenna

The drawing of Figure 5 shows an oblique view of the antenna. In reality, the two dipoles are horizontal, and at right angles to one another. Looking up at the antenna from directly below it, you should see an "X" over your head. Each dipole section is bolted to a mounting plate made up of a foot square piece of heavy plywood. The sections are held to the plywood by means of U-bolts, nuts and washers. The space between the inner tips of the dipoles is not critical, and two or three inches will permit you to make proper connections to the sections.

The mounting plate can be held atop your mast by means of angle brackets, making sure that the brackets and mast do not touch either the dipoles, U-bolts, or line segment A. The dipoles, then, are in a horizontal position and are at right angles to each other for proper operation of the Turnstile antenna. The oblique view shown in Figure 5 assumes that you are looking down on the antenna from above, so don't mount the dipoles one above the other!

Using the Turnstile Antenna

The Turnstile antenna is nondirectional and provides good rejection to automobile ignition interference, which is vertically polarized. It will work well with most CB equipment designed for use with a 50 ohm transmission line system, as the higher impedance line does not adversely affect the operation of the equipment in most instances. It is a pity that the standard 50 ohm SWR meter will not provide meaningful measurements with this antenna,and an SWR meter designed for a 70 ohm line must be used. With the proper instrument an SWR reading of about 1.4 will be noted across all CB channels.

Build A Bi-Square Beam Antenna

Often the need arises for a high gain, inexpensive beam antenna having a bidirectional pattern (a "figure-8" pattern *through* the array with lobes in opposite directions). An antenna of this type is very useful when a CB station is located near a main highway and wishes to radiate a strong signal up and down the highway in both directions. Yagi and Quad type beams are unidirectional antennas that must be turned back and forth to cover two directions. This is a nuisance, and takes time. The Bi-Square beam antenna, on the other hand, covers two directions simultaneously while providing an overall power gain in both directions of about 5.5 decibels over a ground plane antenna. The beam is vertically polarized and works well when communicating with vehicles having whip antennas. Best of all, this beam is simple to build, erect and adjust and is modest in price.

The layout of the Bi-square beam is shown in Figure 6. Basically, it is a single, oversize Quad loop, one-half wavelength on a leg mounted in a vertical plane. The loop is open at the sides, and is fed with a simple two wire matching transformer at one of the open sides. The transformer, in turn, is connected to a random length of RG-58A/U coaxial transmission line running to your station equipment.

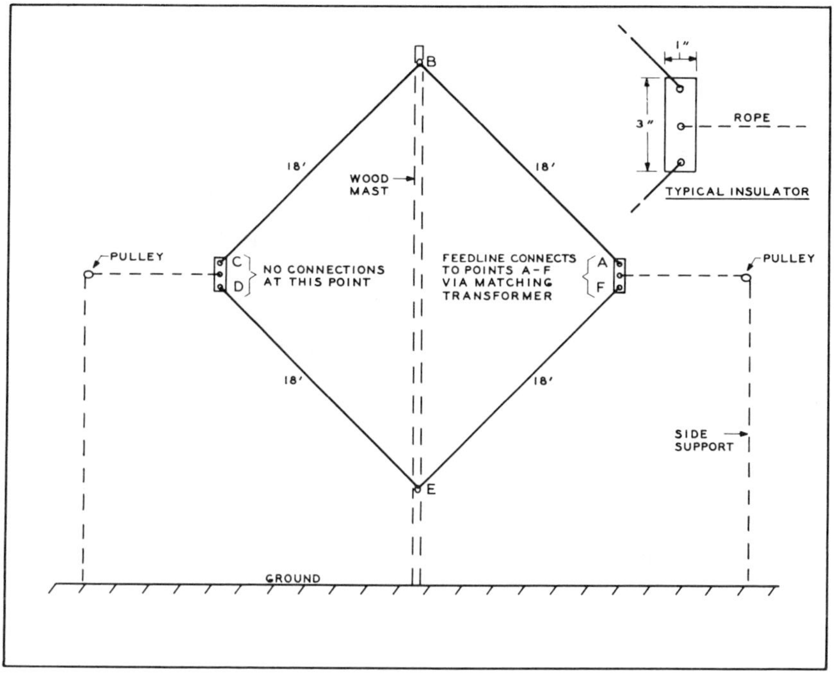

Fig. 6 BI-SQUARE BEAM antenna provides high gain in two directions at right angles to the plane of the wires. Antenna shown is vertically polarized. For horizontal polarization, entire array is rotated 90 degrees so that feed point A-F is moved to the bottom of the mast at point E.

Antenna Assembly

The Bi-square beam antenna is hung from a light weight wood mast consisting of a pair of twelve foot 2X2's spaced with 2X2 blocks. A single twelve foot 2X2 serves as a top section, as shown in Figure 7. Overall height of the wood mast is about 22 feet. The mast is bolted to a short 2X4 support post which is sunk in the ground. With one bolt in place to act as a hinge, the mast may be pushed up into a vertical position and the second bolt put in place to secure the mounting. The side points of the Bi-Square beam are tied off to short masts or to handy nearby trees or structures. Front and back nylon guys can be placed on the top section if it tends to weave about in heavy winds.

The beam itself is a single square formed of two wires, each 36 feet long overall. One wire (A-B-C in Figure 6) forms the top section of the beam; and the second wire (D-E-F) forms the bottom section. A single insulator supports the apex of the array at point B and a second in-

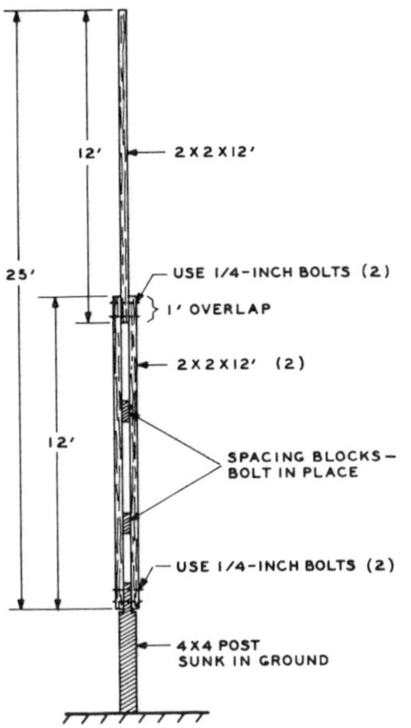

Fig. 7 INEXPENSIVE WOOD MAST used with Bi-Square array may be adapted for other CB antennas. Mast is guyed at the top and also at the overlap point. Mounting post is notched at the top to fit space between the mast uprights. The mast is light enough to be pivoted at the bottom and pushed into an upright position by one muscular CB operator. Mast should be given two coats of outdoor paint to protect it from weathering.

sulator takes up the slack of the bottom section at point E. Inexpensive insulators made out of a strip of polystyrene, plastic or micarta are used at points C-D and A-F. Small aluminum pulleys having nylon rollers are used at the side points of the beam to pull the array into shape, as indicated in the drawing. Nylon line available at most large hardware stores in 100 foot hanks is employed for all guy ropes.

Insulators B and E are attached to the center wood pole and the feed system shown in Figure 8 is attached to the beam at points A-F. Once the pole is erected, the outer points of the beam are adjusted until the antenna is approximately square in shape.

The Feed System

A simple matching system is used with the Bi-Square beam antenna to adapt the antenna to a 50 ohm coaxial transmission line. The matching transformer is made up of two bare #16 copper wires about 9'6" long. Spacing between the wires is not critical and can be 2 to 4 inches. One or two spacers are made of the same material used for the antenna insulators and are placed along the transformer a few feet apart. One end of the two wire transformer is soldered to points A and F of the antenna and the transformer runs inward towards the center support pole. The RG-58A/U coaxial transmission line is brought up the pole and outwards towards the matching transformer and attached to it by means of a movable insulator which slides along the wires and makes contact to them by means of two Fahnstock clips. The end of the coaxial line is prepared with a pig-tail as described in a previous section and the two leads are attached to the clips. Enough slack should be left in the line so that the insulator may be moved back and forth for a foot or two at the end of the matching transformer.

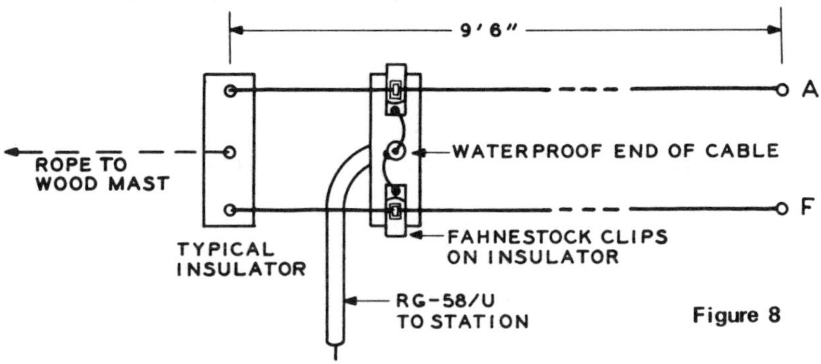

Figure 8

Tuning the Antenna

The Bi-Square beam is bidirectional with maximum signal radiation and reception at right angles to the plane of the wire loop. The loop therefore is positioned broadside to the desired directions. Antenna adjustment takes only a few moments, with the aid of a helper and an SWR meter located at the station. Power is applied to the antenna from the CB transmitter, preferably on a mid-frequency channel (say, channel 11). The tuning expert mounts a step ladder or other support to reach the sliding insulator at the end of the matching transformer. While the assistant watches the reverse reading of the SWR meter, the expert on the ladder grasps the movable insulator and slowly moves it along the transformer section to the point of minimum SWR. Once this point is found, the transmitter is turned off and the clips are soldered to the wires to make a permanent connection between coaxial line and matching transformer. The joints are then covered with waterproofing compound and a coat of vinyl tape. The last step is to trim off the lengths of wire extending beyond the connector. Properly adjusted, the Bi-Square beam should exhibit an SWR of less than 1.5 across all CB channels.

Chapter 14

Build a Quad or Yagi Beam for CB

(Build Your Own Beam Antenna and Save Money!)

You can build your own high gain Quad or Yagi beam and save money over a purchased product. This chapter includes complete information for building four high gain beam antennas that are good projects for the advanced CBer. The first project, a single element Demi-Quad antenna, is also a great beginner's project as it is inexpensive to build and simple to get working. The 2 element Quad and multi-element Yagi antennas are bigger, better and more complex, but still good projects for those CBers who have had experience erecting antennas and scrambling over rooftops.

Many CBers have turned to horizontal polarization in an effort to reduce automobile ignition noise and to reduce the strength of unwanted local signals that have vertical polarization. Shown in this chapter are 3 element, 5 element and 7 element horizontally polarized Yagi beams that combine maximum power gain and low wind resistance in an inexpensive design for the experimenter wishing to try horizontal polarization. The power gain of these antennas can be computed from the information given in Chapters 7 and 8 of this Handbook and by reference to the Truth Table in Chapter 8.

Build a Demi-Quad Antenna

The Demi-Quad antenna is a compact and inexpensive array that may be supported by a single pole or mast. It is light, unobtrusive, and has the same "figure-8" bidirectional pattern through the loop as exhibited by a dipole antenna. The Demi-Quad may be wired to have either horizontal or vertical polarization by proper placement of the feedline, as

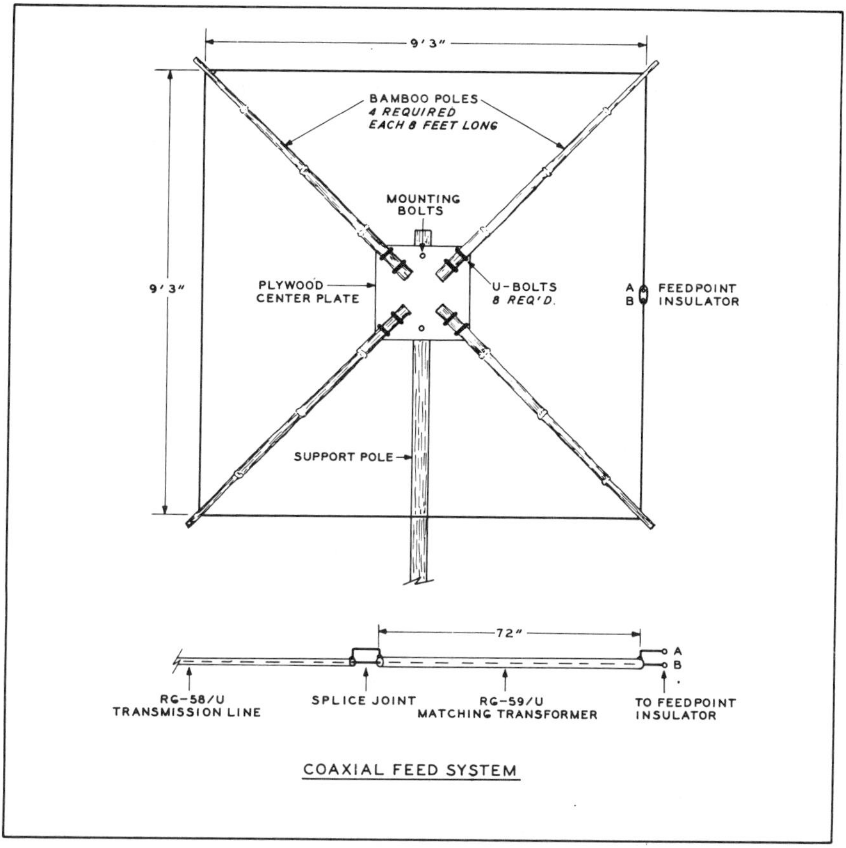

Fig. 1 THE DEMI-QUAD BEAM ANTENNA. This simple one element beam provides worthwhile gain and a bidirectional pattern. Radiation pattern is at right angles to the plane of the loop (into and out of the page). Beam is fed at midpoint of one side for vertical polarization (A-B). Short section of 70 ohm line (RG-59/U) matches 50 ohm line to beam and provides low value of SWR.

discussed later. This antenna may be considered to be a very simple beam having about 2 decibels power gain over a dipole. It requires only a half-turn for complete coverage of the compass and is light enough to be supported by a TV rotator.

Demi-Quad Assembly

The complete Demi-Quad loop antenna is shown in Figure 1. A light bamboo frame is used to support the wire loop in a vertical plane. Each side of the loop is about one-quarter wavelength long, and the loop is broken at one point for the 50 ohm coaxial transmission line and a simple matching transformer made of 70 ohm coaxial line. If the line and transformer are attached to the loop at the middle of the bottom section, the antenna is horizontally polarized. If the connection is made at the middle of one side (either side), the antenna is vertically polarized.

The framework of the antenna is assembled from four lengths of bamboo attached at their large ends to a plywood center plate by means of galvanized iron U-bolts. Each pole is wrapped with vinyl plastic tape between joints to enhance the strength of the assembly and to retard splitting of the bamboo. In addition, each pole is given several coats of waterproof varnish after wrapping. Small holes are drilled as shown near the tips of each pole to pass the antenna wire which is threaded through the poles after assembly of the framework. Each end of the wire is cleaned and the ends are passed through the holes of the center insulator, wrapped back upon themselves and soldered. Enough tension may be imparted to the wires to keep them taut by loosening the center U-bolts and spreading the butt ends of the poles.

The Feedline

The 50 ohm coaxial transmission line must be long enough to reach from the CB equipment to the center plate of the antenna. The recommended line is RG-58A/U. At the antenna end of the line, connection is made to the antenna terminals through a short section of 70 ohm coaxial line (RG-59/U) which acts as a matching transformer. This device matches the 50 ohm line more closely to the radiation resistance of the Demi-Quad antenna, which is about 120 ohms. It is only necessary to cut the 70 ohm line to the proper length for it to make this impedance transformation--more about this later.

Antenna Construction

The first job is to cut the center plate out of 1/2-inch thick plywood and give it several coats of outside house paint. Pay special attention to the edges of the plywood, as water will attack the glue if the edge is not well painted. Drill the plate for the U-bolts and temporarily assemble the bamboo poles to the plate. Mark the exact center of the plate and measure out 6'6½'' on each arm *from the center point.* Mark the distance on each pole. This is the spot where you will drill a hole in each arm through which to pass the antenna wire. Stretch the wire out into a straight line, and temporarily attach it at the marks by means of a piece of string and some tape. It should be a tight fit. If it seems too loose, you may have to mark a new hole a little farther out on one or two poles than the position you have just marked. When you have found the correct points, drill each pole carefully with a drill just slightly larger than the wire size. Now, pass the wire through the holes, attach the center insulator at the bottom of the antenna and tighten things up by pushing the butt ends of the poles out a bit in the U-bolts. Don't change the wire length; that determines antenna resonance.

Antenna Polarization

Antenna polarization depends on how you mount the Demi-Quad to the support pole. For vertical polarization, the center insulator should be at the side of the antenna. For horizontal polarization, the insulator should be at the bottom of the antenna. You can experiment with polarization, if you wish, by swinging the antenna from a horizontal to a vertical position, achieving both vertical and horizontal polarization at the same time when the bamboo poles form a vertical cross and all wires are tilted at a 45 degree angle.

The Coaxial Feed System

As mentioned before, the feedline consists of a 50 ohm transmission line plus a special short section of 70 ohm line that acts as a form of matching transformer. The 70 ohm line should be cut to a length of 6'3", which allows 1½ inches at each end to make connections. The line ends are prepared with pig-tails, as described in Figure 3, Chapter 13. Take care in making the splice as it is easy to melt the polyethylene center insulation of the line with the heat of the soldering gun. To make the splice, twist the center conductors together and solder them carefully. When cool, wrap the joint with vinyl tape, continuing the tape over the polyethylene insulation at each end of the splice. The braid pig-tails are now carefully twisted together and soldered. The last step is to wrap the completed joint with two layers of vinyl tape, overlapping the windings as you go to make the wrapping waterproof. A liberal coating of waterproofing RTV-102 compound completes the joint.

You are now ready to attach the free end of the 70 ohm coaxial line to the antenna at the center insulator. The line should be supported from the center pole so that the weight of the line does not pull at the antenna joint. Tape the line to the pole and center plate. The center conductor of the line is soldered to one end of the antenna loop and the outer conductor pig-tail to the other end of the loop. The pig-tail should take the strain if the line is pulled. When connections are completed, the end of the line is wrapped and waterproofed as described in Chapter 9.

Antenna Installation

The Demi-Quad antenna is quite light and may be handled by one small CBer even though the wires have a devilish tendency to tangle with nearby objects! If a metal support pipe or mast is used, take care that the coaxial line and bottom of the antenna do not bang against the metal.

Up she goes! The antenna may be turned by hand (the "Armstrong" method) or by an inexpensive TV rotator. Since the pattern of the antenna is bidirectional and very broad, it is really only necessary to turn the antenna about 180 degrees to obtain complete coverage. Two ropes attached to the lower crossarms will easily turn the antenna and also hold it in position so that it will not be turned by the wind.

The Demi-Quad operates on all CB channels. The SWR may be measured with the usual 50 ohm SWR meter and readings will run between 1.3 and 1.7 across the channels. The flat surface of the loop is aimed in the direction you wish to receive or transmit.

Build A Two Element CB Quad Antenna

As shown in the Truth Table of Chapter 8, the 2 element Quad beam antenna provides a power gain of about 8.8 decibels over the conventional ground plane antenna. This is certainly a worthwhile power gain and is useful on both receiving and transmitting. The 2 element Quad is a good building project for the home constructor who is making his first beam antenna. Quad kits may be purchased and the good ones provide satisfactory results but at a higher price than an equivalent homemade Quad.

The simplest and least expensive Quad assembly is constructed of bamboo support arms and a wood support plate, as described in the previous section. The 2 element Quad is made up of two Quad loops, one serving as the driven element and the other as a parasitic reflector, as shown in Figure 2. The directivity of the beam is from the reflector *through* the driven element. The loops are supported by means of a short horizontal boom which may be made of either wood or metal. Each loop is assembled in the manner described for the Demi-Quad antenna.

The 2 element Quad is a true beam antenna and has a unidirectional (one way) pattern, showing a good front-to-back signal ratio. It is therefore necessary to rotate the antenna through 360 degrees to obtain com-

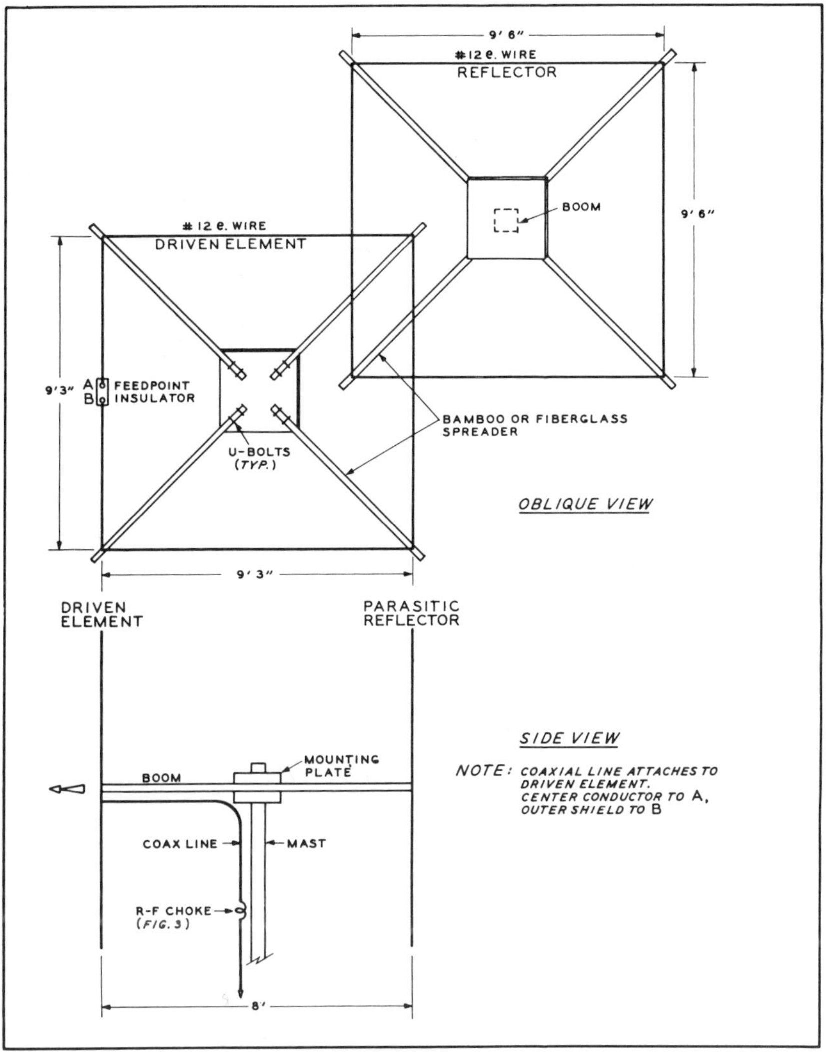

Fig. 2 **TWO ELEMENT QUAD ANTENNA** provides good power gain for CB service. Top, oblique view shows placement of driven element and reflector. Driven element is fed at points A-B. Shield of RG-58/U coax line is attached to point B and center conductor attached to point A. Joint must be thoroughly waterproofed. Reflector is continuous loop of wire with no connection to either feedline or driven element. Side view of Quad is shown below with elements mounted on wooden boom. Beam pattern is to the left. Coax line comes up mast and out boom to driven element. See Fig. 3 for choke data.

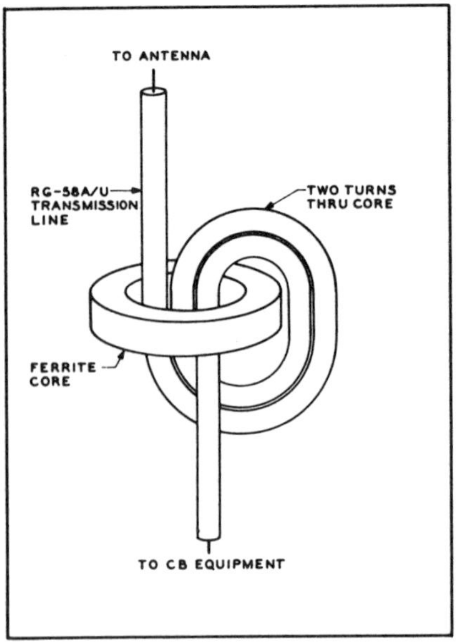

Fig. 3 FERRITE R-F CHOKE COIL. Choke coil is made by winding two turns of coaxial line about a ferrite core. Coil is about two inches in diameter and is placed about a quarter-wavelength down the line from the antenna terminals. Turns may be taped to core. The core is Indiana General CF-111. For data on this core, catalog and list of distributors write: Indiana General Corp., Crow Mills Rd., Keasby, N.J. 08832. Core is Q-1 material,

plete radio coverage. A heavy duty TV rotator will do the job. A photo of a typical Quad installation is shown on page 20.

Antenna Assembly

The Quad loops are assembled in the manner described for the Demi-Quad. Note that the reflector loop is somewhat larger in size than the director loop and the mounting holes for the antenna wires are drilled a bit farther out on the bamboo poles. It is suggested that extra-length poles be used so that the small tips may be cut off and discarded. Bamboo poles, by the way may often be purchased at bamboo distributors in large cities, at some rug stores in smaller towns and sometimes at garden nurseries and hardware stores.

The boom should be made of a section of dry 2X2 lumber, well painted to protect it from moisture in the air. "Green" lumber tends to warp as it dries out, imparting an unlovely twist to the symmetrical Quad antenna. Use dry lumber, sand it well and give it two coats of outdoor house paint.

The center plates of the Quad loops are attached to the ends of the wood boom by means of four galvanized steel angle brackets. The brackets are mounted slightly off-center on the boom so that the retaining bolts will not interfere with each other passing through the boom. Do not use wood screws at these joints as they probably will work loose in the first wind storm. The completed wood and bamboo structure seems to have about as much structural strength as a jellyfish. However, once the wires are strung in position and made taut, the assembly magically becomes strong and amazingly rigid. Believe it or not!

Note that the reflector element has no center insulator and is just a complete circle of wire. The driven loop, as in the Demi-Quad, is broken by an insulator at the feed point. Positioning this loop will determine whether the beam is horizontally or vertically polarized, as described in the previous section.

The final assembly operation is to bolt the loop assemblies to the boom. A little pre-planning at this point is helpful, because once the Quad is assembled it becomes an unwieldy object. A good idea is to place the boom atop a six-foot step ladder to keep the whole antenna in the clear above ground during the final assembly stages. Do not lift the Quad by the bamboo arms as this tends to warp the assembly.

Once the beam is completed, the 50 ohm transmission line should be attached. The line is connected directly to the antenna ends at the center insulator of the driven loop. No matching transformer is required. Previous remarks about waterproofing the end of the line should be remembered at this stage of the game.

The Ferrite R-F Choke Coil

The 2 element Quad has a good front-to-back ratio and good gain. In order to make sure that the presence of the feedline does not upset the electrical characteristics of the Quad, an r-f choke coil is placed in the coaxial feedline. This device prevents r-f energy at the antenna from passing down the *outer surface* of the coaxial line. Remember: all energy must remain *inside* the line. To make a suitable choke coil, the coaxial line is merely passed twice through the center hole of a small ferrite core, making two loops of the line about 2 inches in diameter as shown in Figure 3. The choke coil is positioned about 8 feet down the line from the antenna. After looping the line through the core, tape the coils to the core so that the turns will not move about. It is a good idea to fill the inside of the core with cloth and then tape the whole assembly as the ferrite material is quite fragile, and may shatter if it bangs against the mast on windy days.

Antenna Installation

Normally, you'll build the Quad antenna for vertical polarization with the feedline brought away at one side of the driven element loop. The

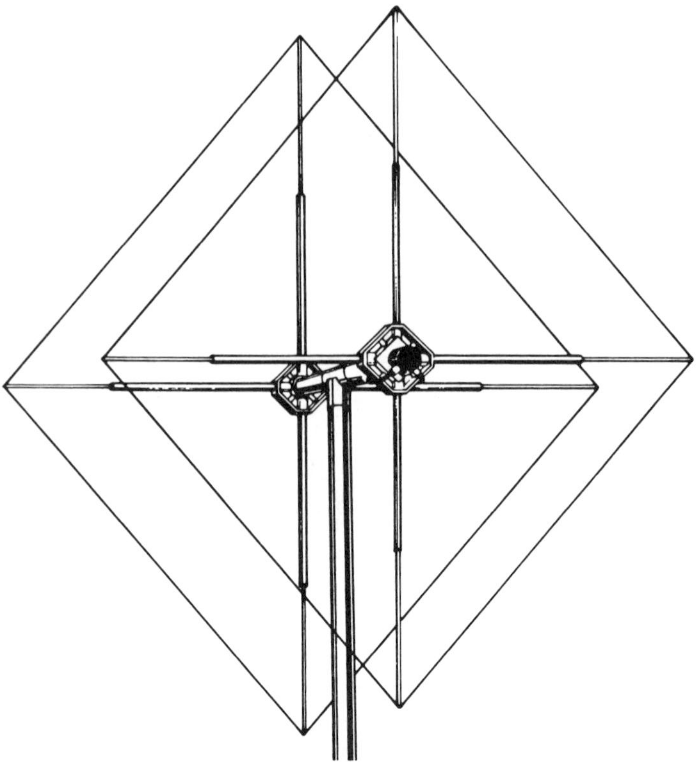

TYPICAL QUAD BEAM ANTENNA. Popular two element CB Quad is made of aluminum supporting boom with rugged center clamp to affix it to the vertical support mast. Clamps for crossarms are heavy, formed plastic material. Arms are aluminum with phenolic tip sections to insulate wires from the metallic structure. The whole design is light and sturdy to withstand heavy winds and gusts which can break the crossarms if they are not properly stressed.

line should be dressed inwards towards the boom, then along the boom to the center support, and then dropped down the mast to the station. Tape the line to the boom to keep it from flopping around in the wind. If you use a TV rotator, you'll have to leave enough slack in the line so that the rotor will not twist the line into a knot when the antenna turns or be stopped by too short a length of cable. Mount the Quad so that the lower wires of the loop are at least ten feet above any structure, such as a roof. As with any antenna, best results will be obtained with the Quad as high in the air as legally possible, and clear of other objects. The SWR on the transmission line will run between 1.2 and 1.6 across the CB channels when the antenna is in the clear.

Build A 3 Element Horizontal Yagi Beam

Many CB operators are switching to horizontal polarization to reduce interference from nearby CB stations and also to drop automobile ignition noise on reception. Some manufactured Yagi beam antennas can be switched over to horizontal polarization by merely rotating the antenna in its mounting fixture. Other beams are not designed for horizontal operation and require reworking before they may be flopped over to a horizontal position.

Described in this section is an inexpensive and effective 3 element Yagi beam antenna designed for horizontal mounting. It is inexpensively constructed and may be turned with a heavy duty TV rotator. Mounted in the clear this beam will provide excellent communication performance for CB service.

Antenna Assembly

The horizontally polarized 3 element beam is shown in Figure 4. It is designed for optimum operation on any CB channel in the 27 MHz range. The antenna consists of a driven element (dipole D) fed at the center with a balancing unit (balun) and matching transformer (segments A and B, Figure 5). These are attached to a 50 ohm coaxial transmission line. Two parasitic elements are used in the beam, a director element (E) and a reflector element (F). The physical length of all elements is pre-set to provide optimum power gain and good signal rejection from the back and sides of the beam. All elements are made of sections of aluminum tubing. The elements are supported by a boom made of a length of 2X2 lumber. The antenna is mounted with the elements parallel to the ground and the beam pattern is in line with the boom through the director element.

Element Construction

The reflector and director elements are made first. As the aluminum tubing is manufactured in twelve foot pieces it is necessary to extend the tips of the tubing to reach the proper length. The tubes are cleaned of dirt and scale and short inserts of clean telescoping aluminum tubing are thrust in each end of the elements and the exposed portion of the insert is varied until the correct over-all length is obtained. Make sure that equal portions of the inserts are exposed to preserve the symmetry of the element. If the fit of the tubing is "sloppy" it will be necessary to slip thin shims of scrap aluminum between the tubes until a firm

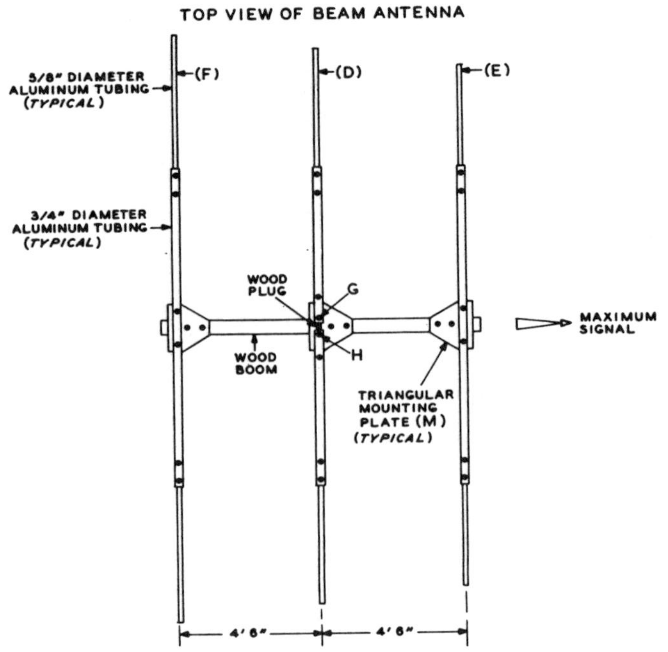

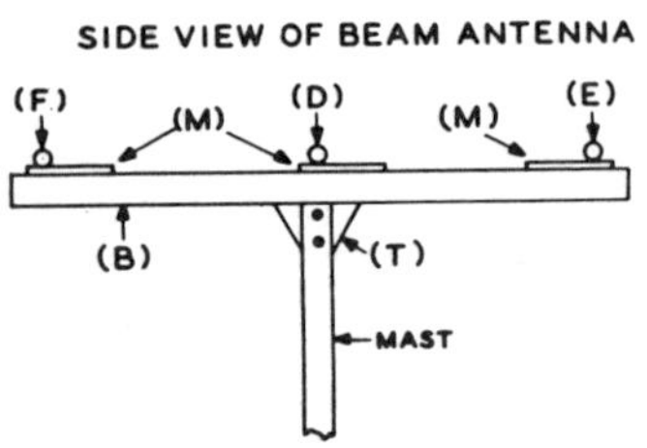

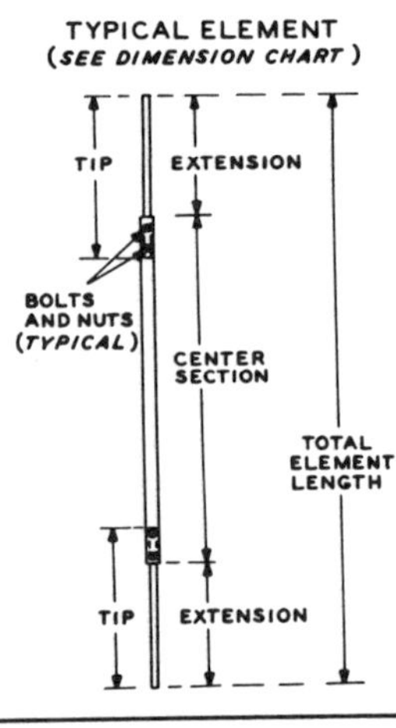

ELEMENT DIMENSIONS

NOTE: *SPACING AT CENTER OF DIPOLE IS ONE INCH*

ELEMENT	CENTER SECTION	TIP	EXTENSION	TOTAL ELEMENT LENGTH
REFLECTOR (F)	10'	4'9"	4'3"	18' 6"
DIPOLE (D)	10'	4'3"	3'9"	17' 7"
DIRECTOR (E)	10'	3'8½"	3' 2½"	16' 5"

Fig. 4 THREE ELEMENT HORIZONTALLY POLARIZED BEAM for CB. Elements are parallel to the ground and supported on wood boom about 9'10''long. Feedline is attached to driven element (D) which is split at the middle. The halves of driven element are driven onto wood dowel for rigidity. Elements are mounted to boom by means of U-bolts and plywood plates. Feed system is shown in Fig. 5.

joint is achieved. Three holes are now drilled through each joint which is pinned in position by galvanized nuts and bolts passed through the holes. As a final step wrap the joints with a layer of vinyl tape to protect them from moisture.

The driven element is made in the same manner except that when completed it is cut in half and the two sections are driven onto a round plug of wood 14 inches long which acts as an insulator and support for the two halves of the element. Dowel rod can be obtained in many large hardware stores or a section of broomstick whittled to size may be used. The two halves of the dipole are pinned to the plug by means of galvanized nuts and bolts passed through holes drilled in the center assembly. The exposed portion of the plug is given several coats of waterproof varnish, and all elements are given a coat of aluminum paint to retard attack of the metal by salts and moisture in the atmosphere.

The Framework

The supporting structure (B) of the antenna is made of a nine foot length of 2X2 lumber. Choose a section that is dry, straight, and free of knots or splits. Sand the wood and give it two coats of waterproof varnish before you start assembly. The elements are held to the boom by means of triangular mounting plates (M) made of plywood. The plates are cut fourteen inches on a side and are given two coats of waterproof varnish. Take care to seal the edges of the plate with varnish to prevent moisture from entering between the layers of wood. The plates are bolted to the center of each element using galvanized hardware. In the case of the dipole, the center bolts holding the wooden plug may also be used to mount the element to the triangular plate. The plates in turn are bolted to the wooden boom. Reflector and director elements are mounted at the ends of the boom and the dipole element is placed in the center of the structure. When the assembly is completed, crush the threads on the ends of the bolts with a hammer to prevent the nuts from working loose in a heavy wind.

The foregoing operations should take place in an open yard or driveway where flat space is available to lay out the boom and the elements in their proper relation to one another. Use care in your assembly to make sure that all elements lie in the same plane so that the finished beam will present a neat, workmanlike appearance.

The Feedline and Matching Network

After the antenna is assembled it is time to turn attention to the feed

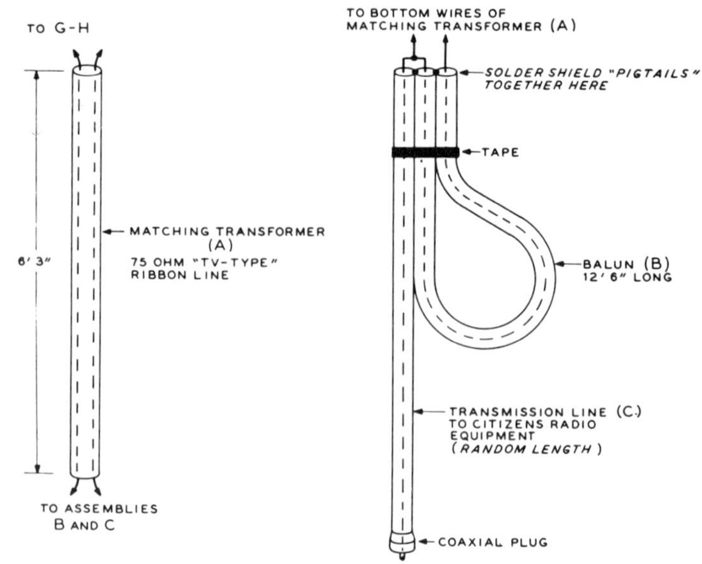

Fig. 5 MATCHING TRANSFORMER AND BALUN. Transformer (A) is attached to center terminals (G,H) of driven element (D) of Figure 4. Other end attaches to wires of balun and transmission line (B). Shields of balun and line are connected together, but no connection is made between shields and (A), or between shields and antenna. Balun and line are RG-58/U and matching transformer is TV "ribbon" line (Belden 8222 or equivalent). Tape all connections.

system (Figure 5) consisting of a matching transformer (A), a balancing unit (B), and the coaxial transmission line (C). Items A and B insure that efficient energy transfer takes place between the transmission line and the antenna. Assembly of these items is a simple task if done properly.

The Matching Transformer. (A). This transformer consists of a 6'6" length of two wire 75 ohm TV-type ribbon transmission line. Each end of the line is stripped of insulation for a distance of 1½" inches and the four exposed wires are tinned with a soldering iron. The two wires at one end of the line are attached to the center bolts (G and H) of the dipole element (D). *Care should be taken that a good electrical connection exists between the bolts and the aluminum tube.* All paint is removed from the tubes beneath the bolt and nut, and the surfaces of the tubes are thoroughly cleaned with sandpaper. When completed, the connections are wrapped with vinyl tape. The opposite end of the transformer is connected to assemblies B and C.

The Balancing Unit (B). The balancing (balun) transformer is made of a 12'6" section of 52 ohm coaxial line. Three inches of the outer vinyl insulation is removed from each end of the line and the exposed braid is made into pig-tails as discussed in previous sections. Next, the inner jacket is removed from the center conductor for a distance of two inches. One end of the 52 ohm transmission line is prepared in a similar manner.

Assembling the Units. The balun is coiled into a roll about a foot in diameter, with the ends laying close to each other. It is taped to hold it in this position. Next, each end of the inner conductor of the balun is soldered to one of the wires of the matching transformer (A). The inner conductor of the transmission line (C) is now soldered to *one wire* (either one) of the matching transformer. These joints are wrapped with vinyl tape. The ends of the coaxial lines are also taped. The last step is to twist all three pig-tails together, solder them securely and trim them short. Tape this joint to prevent the pig-tails from touching the other wires or connections.

Erecting the Antenna

The antenna is now ready to be placed in the operating position. If a TV mast and rotator are used, the boom of the antenna can be bolted to a triangular plywood mounting plate (T) which in turn is clamped to the supporting pipe by means of TV-type U-bolts. To prevent the weight of the balun and transmission line from damaging the matching transformer the balun should be taped to the boom of the antenna directly below the dipole, allowing the matching transformer (A) to swing in a loop beneath the antenna. The weight of the transmission line may be taken care of by fastening the line to the tower with tape or string. Enough slack should be left in the line so that the antenna may be rotated through a full circle without fouling the cable.

Antenna Maintenance

The beam antenna requires little care or maintenance once it has been erected in position. If there is a high degree of industrial fumes, salt spray, or moisture in the atmosphere it will be necessary to lower the antenna and repaint the aluminum elements every year or so. Deterioration of the coaxial transmission line is not a problem as long as the outer sheath of the line is not punctured, and as long as the ends of the line are sealed against moisture.

It must be remembered that an antenna of this size has a fair amount of wind resistance, and large forces work against the elements and the boom in a storm. Past experience with antennas of this type has shown that the antenna will withstand the forces of the winds best when the elements are positioned *into the wind,* that is, the wind blows broadside against the boom.

Using Your Beam Antenna

In use, the beam is aimed in the direction of the station you wish to communicate with. If this direction is unknown it is possible to peak the antenna on the station by merely listening to the station and swinging the beam until maximum signal strength is achieved.

If it is desired to communicate in a certain direction, the boom of the antenna is aimed in that direction with the director forward of the dipole as shown in the illustration. If general coverage operation is desired, some means of rotating the beam antenna is required. A TV-type rotor may be used for this purpose. Some of the better rotors have a compass indicator showing the exact heading of the beam.

A less expensive installation may be made by attaching the beam to the supporting structure by means of two heavy strap hinges mounted in line one above the other. The beam can swing about the support on the hinge mount. Ropes tied to the extremities of the boom are used to hold the antenna in the desired direction. The beam may be turned to a new heading by pulling on one of the ropes.

Proper maintenance of your antenna will result in trouble free operation. Keep the wood parts well painted, and check all metal parts and fittings for rust and corrosion. A coat of aluminum paint will lengthen the life of metal parts.

Build a 5 Element Horizontal Yagi Beam For "Top Man on the Channel" Performance

The 3 element beam described in the previous section can be rebuilt to provide the ultimate in power gain for an array of this size. The 3 element beam provides a power gain of about 8 decibels over a dipole. With the addition of two extra director elements on a long boom, the power gain is boosted to about 11 decibels. This means that the 5 element Yagi beam will make a 5 watt CB transmitter sound like a 60 watt transmitter at the distant point of reception. This gain, too, is also

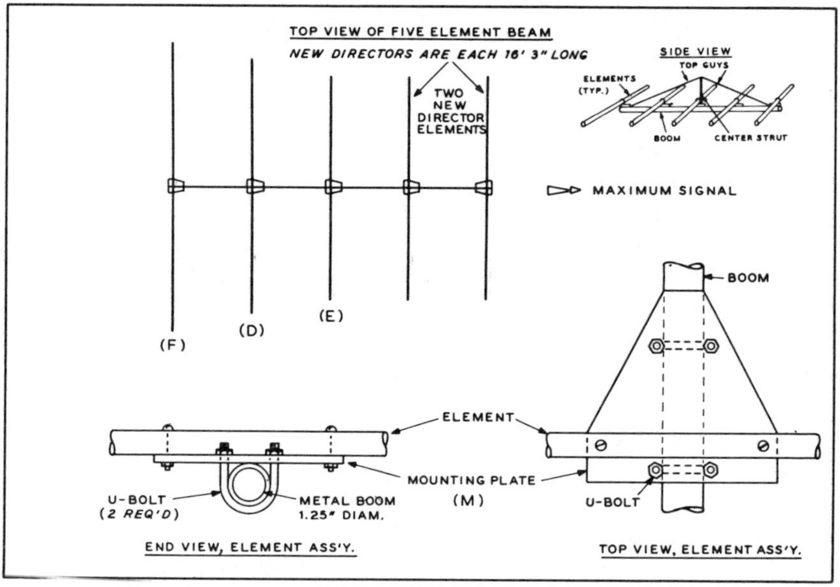

Fig. 6 FIVE ELEMENT HORIZONTAL YAGI provides big signal for CB service. Metal boom is made of TV mast sections, with top guys for added strength. Guys are broken at mid-points with egg insulators. Elements are affixed to boom with plywood mounting plates. Element dimensions shown in Figure 4 are used and spacing is 4'6" between elements. See Figure 5 for matching system.

available for reception. For operation over an extreme legal range, over mountainous terrain, or under conditions of heavy interference, this inexpensive beam antenna will "deliver the goods".

Beam Assembly

The 5 element Yagi beam is constructed on a 20 foot metal boom and consists of a reflector, a driven element and three directors. Element spacing is about 5 feet, and element lengths are given in Figure 6. The use of a metal boom is mandatory because of the weight and leverage of the many elements, and a top guy is added to prevent the antenna from swaying in a heavy wind.

The boom is made of two pieces of 10 foot long, TV-type swaged antenna mast sections, made of seamless, #16 gauge, hot-dipped galvanized steel. The elements are made of lengths of telescoping aluminum tubing fixed to the boom by means of plywood mounting plates. In this design, the elements are insulated from the boom by means of the plywood plates.

The two sections of the boom are slipped together and bolted in position with two #10 galvanized machine bolts, nuts, and washers to form a twenty foot boom. A single top-guy runs the length of the boom to a short center strut, providing vertical bracing to the array. The guy is made of galvanized iron wire and is broken in two places by ceramic strain insulators. The strut is a two foot length of ½-inch water pipe, screwed into a mounting foot bolted to the triangular mounting plate (M) of the center element.

All elements must be insulated from the metal boom. This is accomplished by the plywood mounting plates, each of which is held to the boom by means of two galvanized U-bolts. Make sure the element mounting bolts do not touch the boom and that the U-bolts do not touch the elements and all will be well.

The elements are spaced evenly along the boom, allowing almost five foot separation between the elements. When the assembly is complete it may be balanced by hand to find the exact center of gravity which will fall somewhere near the first director (E). The mounting plate (T) is affixed to the metal boom at this point with the aid of two U-bolts. No harm is done if the metal boom is actually clamped to a metal supporting structure or mast, as the boom has no electrical connection to any active antenna elements.

The matching system and feed line shown in Figure 5 are used directly with the five element beam with no alterations. Since the driven element (D) is mounted off-center, the matching transformer (A) should be allowed to drop down beneath the antenna without touching the metal boom. The junction of the balun (B) and the transmission line may be taped to the supporting structure, allowing the transformer to swing about the tower as the antenna is rotated.

Antenna Installation

As this antenna is fairly large it develops considerable inertia during rotation. In addition, the surface area of the antenna produces a high level of torque at the supporting shaft during a heavy wind. The usual inexpensive "TV-type" rotor will not stand up long under such abuse. A heavy duty rotor, such as the *Cornell Dublier* type *HAM-M* is recommended for use with this array.

Maximum SWR over the CB channels should run between 1.2 and 1.5.

Advanced Beam Antenna Construction

The beam antennas described thus far are relatively simple to build.

More complicated and sturdier arrays require a more sophisticated assembly technique. Then, too, as you -- the experimentally minded CBer -- advance in construction practice, you'll undoubtedly want to build bigger and still better CB antennas. The following discussion covers some advanced beam antenna construction techniques that will assist you in planning and building high quality, long-life CB antennas of all types.

A wealth of construction information on beam antennas of all types is included in the following publications:

"Beam Antenna Handbook", by Orr and Cowan

"All About Cubical Quad Antennas", by Orr and Cowan

These books are published by Radio Publications, Inc., Box 149, Wilton, Conn., 06897 and are recommended to all experimentally minded CBers who are interested in building bigger and more effective beam antennas.

Choice of Materials

The strongest, lightest and easiest beam antenna configuration to build is the all-metal array employing a metal boom, with the elements fastened directly to the boom with as few insulators as possible. Parasitic Yagi-type directors and reflectors may, in fact, be attached directly to a metal boom because there is no r-f potential between the exact center of such an element and the boom.

An all-metal array is remarkably weather resistant when built of the proper materials. In contrast, the insulating materials commonly used in less expensive antenna structures (ceramic, plastic, wood or micarta) may be broken by heavy wind stresses on the array, and may deteriorate from the corrosive action of sun, rain and salt spray.

Outranking other materials, aluminum and steel are employed for the framework and elements of most manufactured and home-built antenna arrays. The elements and supporting boom are usually made of aluminum and the clamps and accessories are made of steel, heavily plated to prevent rust. In some expensive arrays, stainless steel is used for the support structure of a large antenna.

The Elements

Elements for Yagi beams may easily be constructed from aluminum tubing of the proper alloy. A summary of the various alloys of tubing is given in Figure 7. The higher strength alloys are a combination of aluminum and copper or aluminum and zinc. The softer alloys are almost pure aluminum. It is important to note that the deflection of the tubing

ALUMINUM TUBING

(STRENGTH DECREASES ↓; CORROSION RESISTANCE DECREASES ↑)

USE	DESIGNATION
AIRCRAFT	7075
AIRCRAFT	2024
AIRCRAFT	2017
COMMERCIAL	6061
INDUSTRIAL	6063
GAS LINES, ETC.	5052
CONDUIT	3003
——	1100

SUFFIX

0 = ANNEALED
12 = 1/4 HARD
14 = 1/2 HARD
16 = 3/4 HARD
18 = FULL HARD

TYPICAL ELEMENT ASSEMBLY

CENTER SECTION	TIPS
$1\frac{1}{2}$" DIAM., .049" WALL	$1\frac{3}{8}$" DIAM., .035" WALL
$1\frac{3}{8}$" DIAM., .049" WALL	$1\frac{1}{4}$" DIAM., .035" WALL
$1\frac{1}{4}$" DIAM., .049" WALL	$1\frac{1}{8}$" DIAM., .035" WALL
1" DIAM., .035" WALL	$\frac{7}{8}$" DIAM., .035" WALL
$\frac{3}{4}$" DIAM., .035" WALL	$\frac{5}{8}$" DIAM., .028" WALL
$\frac{5}{8}$" DIAM., .035" WALL	$\frac{1}{2}$" DIAM., .028" WALL

NOTE- *CLEARANCE BETWEEN TUBING SIZES IS 0.05" IN ALL CASES*

Fig. 7 ALUMINUM TUBING is available in various alloys. Tubing sag under a given load is independent of alloy. Soft alloys are less susceptible to corrosion than are hard alloys. Recommended alloys are 6063 and 6061.

under a given load is independent of the alloy. Equal tubes of soft or hard aluminum alloy will deflect the same amount under a given load, emphasizing the fact that the element sag in a Yagi beam antenna is a function of tubing diameter and wall thickness, rather than of the strength of the tubing.

The corrosion resistance ability of aluminum tubing, on the other hand, is almost an inverse relationship to the strength of the tubing. Thus, the softer alloys are less susceptible to corrosion than the harder alloys. In general, the commercial alloys of 6063 or 6061 are best to employ for Yagi antenna elements as these grades are strong, relatively inexpensive and readily obtainable. The alloy 2024 may often be obtained in surplus aluminum "junk yards". It is considered to be an aircraft grade alloy and is quite expensive when purchased new. The difference in strength between all of these alloys is relatively minor.

Aluminum is properly classified into *tubing* and *pipe*. Tubing is the more expensive as it is drawn and wall thickness is closely controlled. It normally comes in 12 foot lengths. Pipe is less expensive, has lower tensile properties and is extensively used for irrigation purposes. Both tubing and pipe are manufactured by the Aluminum Company of America, Kaiser Aluminum Co., Reynolds Metals Co., United States Steel Co., and others. Sources of supply may be found by phoning the local sales office of these companies, listed in the Yellow Pages of your telephone

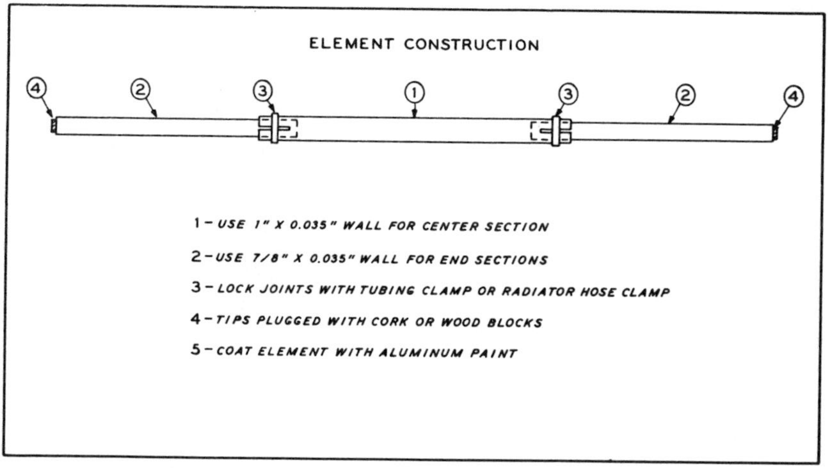

Fig. 8 ELEMENT ASSEMBLY IS EASY when you do it right. Telescoping sections of aluminum tubing are fixed in position with hose clamps as shown here.

directory. In addition, many metal supply houses in the larger cities carry an extensive stock of aluminum tubing.

Aluminum tubing may also be obtained in rectangular and square shape. While more expensive than round tubing, the square or rectangular shape is ideal for an antenna boom, since the element supports can be bolted more easily to the flat surfaces of such a boom. The antenna experimenter is urged to make his boom of square or rectangular material, as antenna changes may be made with ease when this material is used.

How to Assemble Yagi Elements

The first step is to slot the ends of the center section of tubing. A narrow slot about six inches long is cut in each end of the tube with a hacksaw. The slot should go through both walls, on a line through the center axis of the tube. All burrs are then carefully removed from the walls of the tube. The mating section of the tip tubes should be sanded and cleaned to lessen the possibility of seizure after the tubes are telescoped. When pressure is put upon the ends of the center section of tubing, the tip tube should be held firmly in place. The tubes should overlap about six inches.

Before the element is assembled, precautions should be taken to prevent corrosion at the joint. A special antioxidizing compound (*Penetrox A*, manufactured by Burndy Co., Norwalk, Conn. and distributed by the

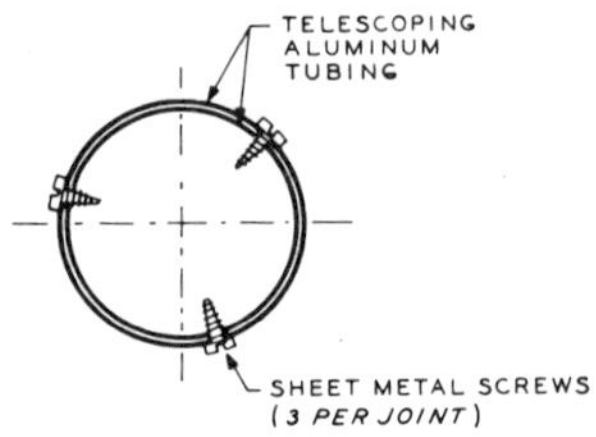

Fig. 9 ELEMENT JOINT is pinned by three self-tapping metal screws in place of tubing clamps. Screws have tendency to loosen in wind unless joint is pinned in three places, about 120 degrees apart, as shown. Drill holes through tubes, somewhat smaller than screw diameter for easy insertion of screw.

General Electric Supply Co.) is smeared lightly over the mating pieces of tubing. When the tubes are telescoped, this compound forms an airtight seal, preventing corrosion. The compound is a good electrical conductor, and provides a trouble-free joint if the other directions are followed carefully. A 5 oz. tube of Penetrox is sufficient to coat all the joints of a four or five element Yagi array.

The element is assembled as shown in Figure 8. The tubing joints may be fastened by means of tubing clamps, hose clamps, or home made circular straps. If the tip sections do not make a close fit with the center section, the joint can be shimmed with small strips of thin aluminum. Self-tapping sheet metal screws can also do a good job if three of them are used at each joint, as shown in Figure 9.

Build a "Long John" Seven Element Yagi for CB

Many CBers have turned to horizontal polarization to reduce interference from vertically polarized stations and also to escape ignition noise interference. The Monster Quad beam shown in the next chapter may be turned into a horizontally polarized array by moving the feed point from the horizontal side to the center of the bottom side of the driven loop.

Some home constructors, however, are wary of constructing such a huge antenna as the Quad and others are firm boosters of the more conventional Yagi beam antenna. Shown in this chapter is a practical "Long John" Yagi beam for CB operation that provides a power gain of over 12 decibels as compared against a dipole antenna. This beam, then, is equivalent in performance to the Monster Quad and is suggested to those CBers who are proponents of the Yagi array. The Long John

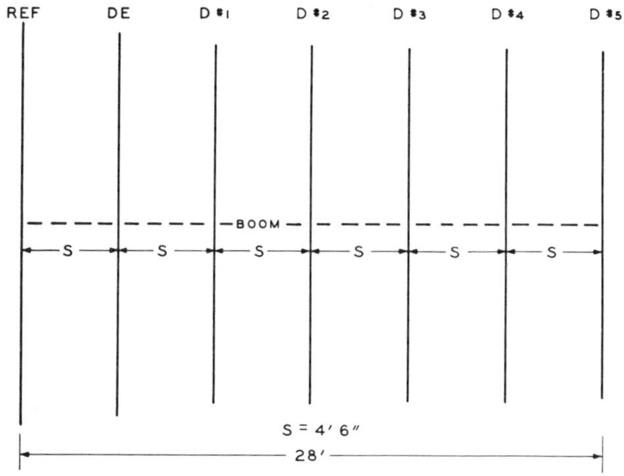

	REF	D.E.	D1, D2	D3, D4	D5
ELEMENT LENGTH	18' 6"	17' 7"	16' 6"	16' 3"	16' 0"

Fig. 10 LONG JOHN YAGI BEAM provides "sock" on all CB channels. This deluxe, horizontally polarized beam (shown top view) has seven elements on a 28 foot boom. Elements are mounted directly to the boom, as shown in Figure 11. Gamma match system is used, as described in Chapter 16. Long John beam has better than 12 decibels power gain over a dipole antenna. The elements are made of 1" diameter tubing with 7/8" diameter tip sections. A square boom made of aluminum extrusion is used to simplify construction. A round boom may be used with the elements held to mounting plates by large U-bolts. The plates, in turn are held to the boom by means of additional U-bolts. A top guy should run the length of the boom, as shown in Figure 6.

beam has seven elements (one reflector, the driven element, and five directors) and has a tight, fifty degree wide pattern.

Beam Construction

A layout drawing of the Long John Yagi beam is shown in Figure 10. It is constructed on a 28 foot boom made of 2½" square x .062" wall, 6061-T66 aluminum tubing. Spacing between the elements is 4'6".

All elements are constructed of telescoping aluminum tubing. The center sections are 1-inch in diameter and have a .035" wall. The tip sections are 7/8-inch in diameter and have a .035" wall. The center sections are 12 feet long and all tip sections are cut 4' long, which allows generous overlap at all joints. The elements are constructed as previously described, and all are bolted directly to the metal boom by means of U-bolts and aluminum mounting plates as shown in Figure 11. The driven element is matched to the 50 ohm coaxial transmission line by means of the Gamma match system, described in Chapter 16 of this Handbook. Only the Gamma match requires adjustment to provide optimum antenna operation, as all elements are pre-cut to proper length.

The Long John antenna is a powerful beam and its construction is somewhat more complex than the run-of-the-mill Yagi antenna. It is recommended to those experienced CBers who have built and adjusted smaller antennas of this type. It is not suggested as a project for the beginner! Those hardy souls who do build a Long John Yagi will be most handsomely rewarded for their effort by the satisfaction of having a truly fine, high gain beam antenna.

After the Gamma match is properly adjusted, the Long John Yagi will show an SWR of less than 2 across all CB channels, with a minimum SWR value close to 1.2 at the frequency of adjustment of the Gamma match.

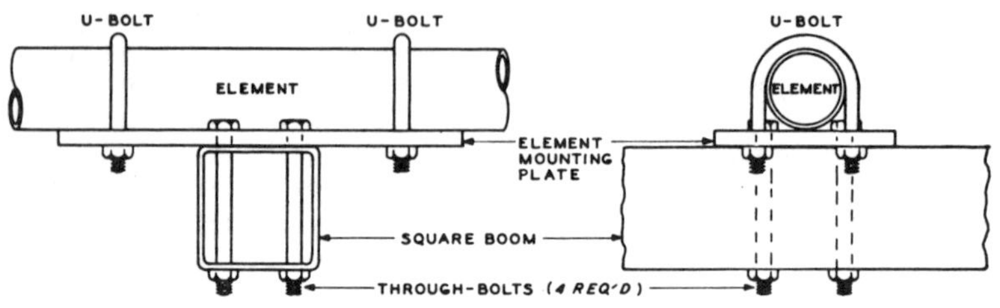

Fig. 11 SQUARE BOOM simplifies beam antenna assembly. Elements are bolted to a mounting plate which, in turn, is bolted to square boom. Boom is drilled for through-bolts, as shown at left. Aluminum boom and mounting plate should be painted with aluminum paint and rust-proof hardware should be used. Lock washers or lock nuts prevent hardware from loosening in heavy wind.

Chapter 15

Build a Monster Quad Beam for CB

(The King of the CB Antennas!)

Would you like to have one of the outstanding signals on your channel? A signal that would insure reliable communication even under conditions of heavy interference? Of course, no antenna can deliver this result 100 percent of the time, but you can approach this outstanding performance by using a Monster Quad beam, the King of CB antennas! The Monster Quad, an impressive and effective 4 element beam, makes your 5 watt input transmitter equivalent to one of nearly 125 watts input working with a ground plane antenna. That is to say, the substitution of the Monster Quad at your station for a ground plane antenna is equal to boosting your transmitter power input to 125 watts in a completely legal manner. Best of all, the impressive power gain of the Monster Quad is also achieved on received signals as well as during your transmissions. The Monster Quad can literally make a weak signal jump up right out of the noise level and interference encountered when using a low gain, nondirectional antenna for receiving.

The Monster Quad also cuts down interference received from signals arriving off the sides and back of the beam, as it provides a narrow, tight 50 degree pattern, as shown in the polar plot of Figure 1.

The Monster Quad beam is a complex antenna assembly and its construction is recommended only to those knowledgeable CBers who have gained experience in building beam antennas and who, in addition, have the ground area to assemble the beam. If you have the know-how and the space, be assured that your efforts and the expense will be amply rewarded by the signal power gain of this superb beam antenna.

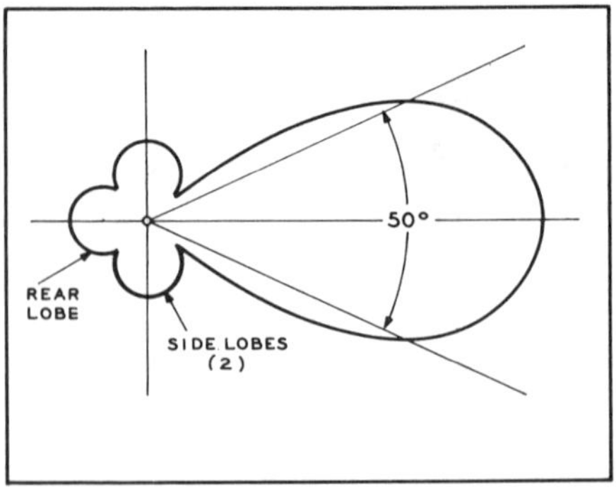

Fig. 1 MONSTER QUAD has tight 50 degree radiation pattern seen from above the array. Reduction of signals from sides and back of the beam pattern is very good. Array must be aimed accurately for best results. Cornell-Dubilier HAM-M rotator is recommended.

Monster Quad Planning

Building a Monster Quad is a real project, but one that can bring you a lot of enjoyment and satisfaction. You are allowed considerable leeway in building such an antenna, the basic electrical circuit of which is shown in Figure 2. Electrically, the Monster Quad is a four element array using two director loops, a driven element loop and a single reflector loop. Each loop is approximately one-quarter wavelength on a side (Figure 2).

Spacing between the loops is not critical and may be as little as 4 feet or as great as 6 feet, with only minor change in beam performance. When the smaller element spacing is used, the power gain of the Monster Quad is decreased about one decibel from maximum value, but the front-to-back ratio is maximum. At the greater element spacing, the power gain is at maximum value at the expense of a slightly decreased front-to-back ratio.

Realistically speaking, spacing really depends upon the overall boom length which the builder wishes to use. Overall boom length is about 12 feet if 4 foot element spacing is chosen, and 18 feet if 6 foot element spacing is used. If rectangular aluminum tubing is chosen for the boom, maximum spacing may be employed, as this type of tubing can be purchased in lengths up to 21 feet. If round tubing is used, maximum tubing length is 12 feet and minimum element spacing must be employed unless the builder has the expertise to splice two tubing sections together. Round aluminum pipe, on the other hand, may be obtained in 20 foot lengths.

Regardless of the shape or length of the boom, it is recommended that the wall thickness be not less than 0.065'' and the cross section not less than 1½ inches. The particular Quad built for these tests uses a 16 foot boom made of a single section of 1½ inch diameter x 0.065'' wall, 6061-T6 aluminum tubing. Spacing between the elements is 5'4''

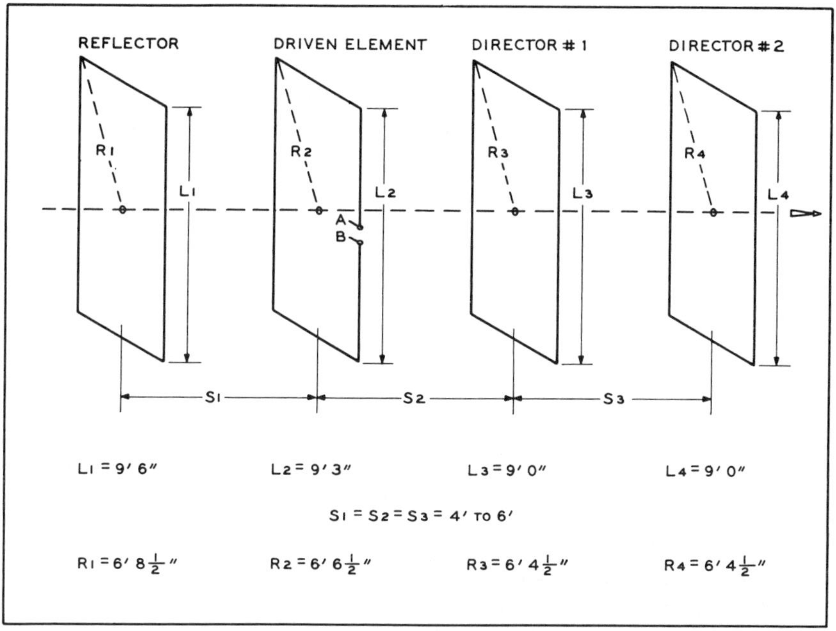

Fig. 2 ELECTRICAL DIAGRAM OF MONSTER QUAD. Reflector and two director loops are used in this effective beam antenna. Driven loop (second from left) is fed by coaxial transmission line at points A-B for vertical polarization. Parasitic loops (reflector and directors) are continuous wires, having no electrical connections made to them. Loops are square, with top and bottom wires parallel to the ground. Side wires are in vertical plane. Loops are made of #14 enameled copper wire. Approximate dimensions from center of loop to corner (R1-R4) is given to aid in drilling the Quad arms.

The builder also has the choice between square or diamond loops. Both loop configurations work equally well and the choice is left to the builder. For convenience in assembly, however, it is a good idea to support the Monster Quad above ground on some kind of wood framework and the use of square loops permits the array to be worked on somewhat nearer the ground than when the diamond shaped loops are used.

The builder also has the choice of horizontal or vertical polarization. Since the reflector and director loops are continuous wires, they work equally well with either polarization of the driven element. The choice of polarization, then, is determined simply by the location of the feed-line point (A-B) on the driven loop element. If the loop is broken for the feedpoint at the bottom, polarization is horizontal. If the loop is broken at the middle of one side, polarization is vertical, as shown.

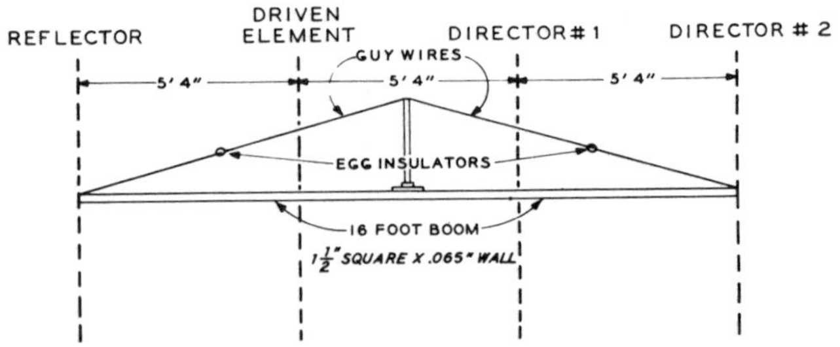

Fig. 3 MECHANICAL ASSEMBLY OF MONSTER QUAD. The Quad elements are mounted on a square boom by means of the plywood plates shown in Figure 4. Top guy wires are used to brace the boom and each guy is broken at the middle by an egg insulator. The center mast is made of a short length of 3/4-inch pipe attached to the boom by means of a pipe flange. Most commercial Quads use round tubing for the boom, but this configuration employs square tubing for greater ease in assembly. Boom is attached to the mast or rotator by means of flange plate at center point.

Monster Quad Assembly

Once the general planning is completed, the builder should study his assembly problem. Because of the weight and wind resistance of the Monster Quad, it is recommended that top guys be used, as shown in Figure 3. A short mast is bolted to the center point of the Quad boom and guy wires run from the top of the mast to each end of the boom. This support provides a rugged framework for the Quad loops and one that will withstand strong gusty winds, icing and other indignities inflicted upon the antenna by inclement weather.

You'll find that several manufacturers provide Quad antenna hardware (such as cross arm supports, Fiberglas Quad arms, etc.). One or two sell complete Quad kits. You can go this route at an increase in antenna cost, or build your own hardware, as suggested in this section. In any case, assemble all your hardware and make sure all parts fit together before you start the final assembly. Refer to the previous chapter, too, for loop assembly information.

It is recommended that the four Quad wire loops be assembled first, followed by the supporting boom structure. The parts are then assembled, with the boom supported on a temporary framework such as two step ladders. Lash the boom to the ladders to prevent it from slipping off.

Waterproof Your Bamboo Quad Arms With Fiberglas

Bamboo arms, even when coated with shellac or varnish and wrapped with vinyl tape, have a relatively short life. Wind, rain and sun tend to shrivel and split the soft bamboo, and sooner or later the arms will warp out of shape or allow the wires of the Quad loop to sag badly. A 2 element Quad can be taken down by one man and repaired. Not so with the Monster Quad, which is a large, heavy array. You can extend the life of the bamboo arms with a protective coat of Fiberglas compound that strengthens them and extends their life. The treatment costs little and takes only an hour or so. This is how you do it:

Materials for Fiberglas treatment can be obtained at Marine hardware stores, large building supply establishments, and some mail order houses. You will need a roll of three inch Fiberglas cloth, and a can of liquid Fiberglas cement. One name for this liquid is *"Boat Resin."* When you buy the liquid you will also get a smaller can of solvent to mix with the resin.

The first step is to lay the bamboo poles out between two boxes or other supports so that the bamboo is clear of the ground. Clean the poles with a damp cloth to remove dust and dirt, and spirally wrap each pole with the cloth until it is completely covered. You can hold the cloth in place with rubber bands or narrow strips of paper "masking tape." The cloth windings should overlap about an inch as you wind the material about the pole. The last step is to mix the solvent with the resin according to the instructions, and give each cloth covered pole a liberal coating of the resin. Use an old paint brush and work smoothly and rapidly from one end of the pole to the other. Let the poles dry over night, then give them a second coat of resin. When the liquid dries, it forms a firm, hard intermixture with the cloth. The treated bamboo pole is now exceedingly strong and you need not worry about the pole cracking or splitting.

For those who want the best, Fiberglas poles for Quad construction may be obtained from various manufacturers.

Raising the Monster Quad

The technique used to place the Monster Quad safely atop the support structure depends upon the ingenuity of the builder and the "lay of the land" around the working site. In particular, if the tower has guy wires or other supports, care must be taken not to entangle the Quad in the guys during the erection process.

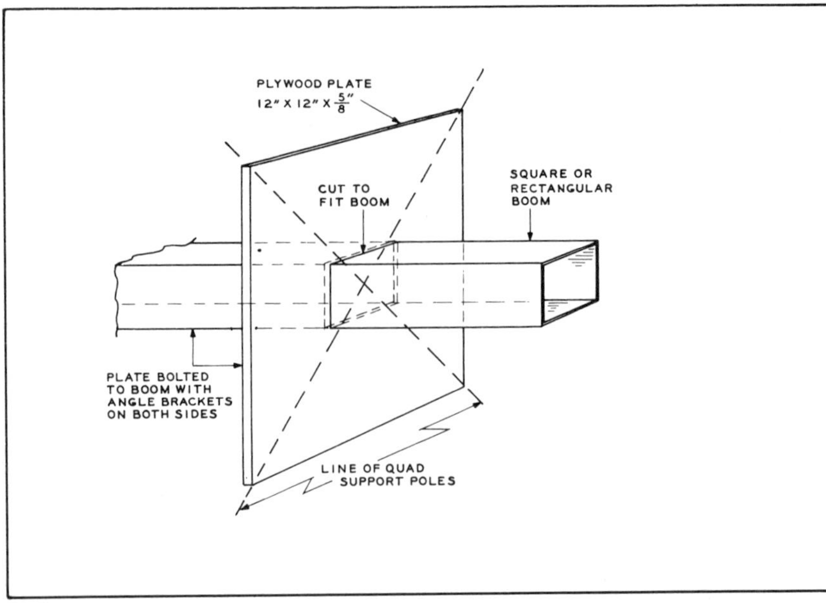

Fig. 4 MOUNTING PLATE FOR MONSTER QUAD. Bamboo arms are attached to a plywood mounting plate affixed to boom with large metal angle brackets.

Some builders disassemble the Quad on the ground and raise it to the top of the tower in pieces and reassemble it there. A less time-consuming task is to lift the Quad at the center of gravity by means of a temporary block and tackle fixed at the tower top. One person atop the tower (assuming it's strong enough!) and a helper on the ground can do the job with relative ease. If it is necessary to disassemble the Quad, it is suggested that the outer reflector and director be removed and the boom and center elements raised in one piece. Once atop the tower, the boom may be temporarily tilted from the horizontal position and the missing elements mounted in place.

Quad Hardware

Homemade Monster Quad hardware designed for a square boom is shown in Figure 4. Wood mounting plates are used with steel hardware. Brackets and small parts are well painted to prevent them from rusting or weathering. Nuts and bolts are galvanized or plated to inhibit rust.

The Monster Quad Feed System

The Monster Quad is designed to be directly fed with a 50 ohm co-

Fig.5 FEED SYSTEM FOR MONSTER Quad. Feedline is coiled around a ferrite core to make r-f choke coil. Construction of the choke is described in previous chapter. It may be fastened to the boom of the antenna by means of electrical tape. Since the ferrite material is fragile, care should be taken to prevent the choke from banging against the boom in heavy winds. Coil may be 2 to 4 inches in diameter (not critical).

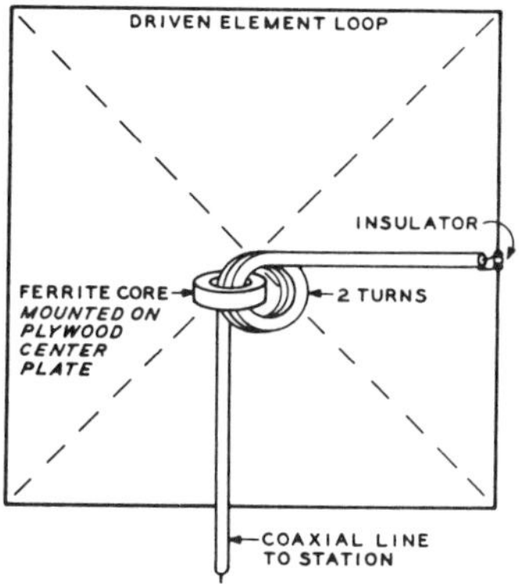

axial transmission line. For a long run of line, heavy duty RG-8A/U cable is recommended. In order to preserve the excellent radiation pattern and good front-to-back ratio of the beam, and to prevent r-f energy from appearing on the outside of the transmission line, it is necessary to place a ferrite core r-f choke coil on the feedline (see chapter 14). The coil is made in the fashion described and placed about 8 feet down the line from the antenna terminals (distance not critical).

The feedline should drop down directly beneath the antenna, after being brought from the driven element (points A-B, Figure 2) to the boom, then along the boom to the vertical support mast. When the line reaches ground or roof level, it may be brought away from the antenna in a horizontal plane, if necessary.

The operator always runs the risk of the feedline becoming a part of the antenna radiating system, and this can happen if the feedline runs parallel to the antenna elements for any length. When the antenna and feedline are electrically coupled together in this fashion, the first symptom that the operator notices is that the front-to-back signal ratio of the antenna is poor or nonexistent. In severe cases of unwanted coupling, the SWR on the transmission line is unusually high.

As in the case of other beam antennas, the user of a Monster Quad should take care that his feedline to the antenna does not pass in the high intensity field in front of the antenna. Dropping the feedline directly down beneath the antenna is good engineering practice and helps to prevent unwanted interaction between line and antenna.

Using Your Monster Quad

After your Monster Quad is atop your tower, you should run an SWR

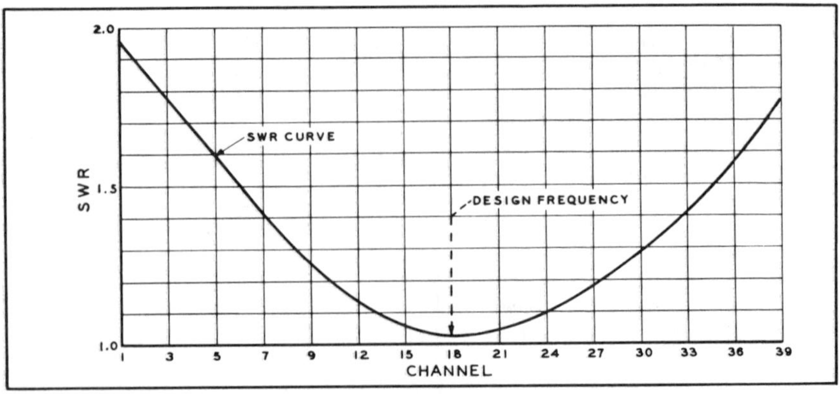

Fig. 6 SWR CURVE OF MONSTER QUAD. This antenna shows an SWR of less than 2 across all CB channels when properly built. RG-58A/U coaxial feedline is used. Larger, RG-8A/U line should be used for long run.

check on it. The SWR across all CB channels should be less than 2, and should approach 1.2 at the mid-channels. But the real proof of performance will be noted in received signals. Swinging the beam back and forth on a received signal should show a sharp drop-off in signal strength twenty-five or thirty degrees on each side of the beam pattern. Signals arriving from the side of the beam should be attenuated to a great degree, as well as signals arriving from the rear. It is imperative that a good, heavy duty rotator, such as the *Cornell-Dubilier HAM-M* unit be used, in conjunction with an accurate direction indicator.

Fig. 7 RF CHOKE COIL is composed of two or three turns of RG-58/U coax feedline wrapped around ferrite core. Coil is 2 to 4 inches in diameter. See Chapter 14 for more data.

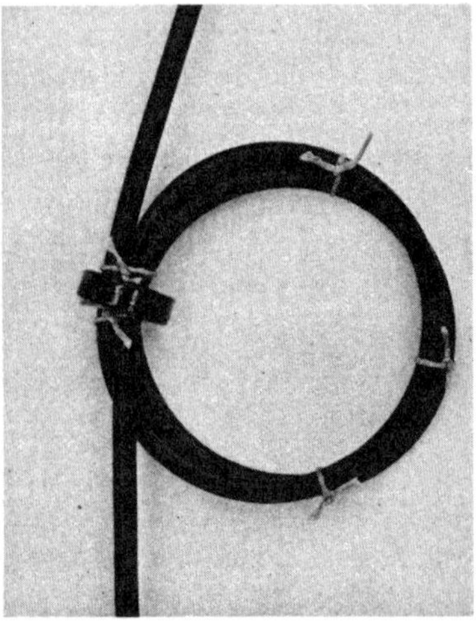

"I swear, the standing waves on my feedline were THIS long!"

Chapter 16

CB Round-Up on Ten Important Subjects

(Baluns, TVI, Lightning Protection-and Lots More)

This chapter covers a number of subjects of great importance to the serious CB operator which do not deal directly with antennas. However, unless your antenna is used properly and in a safe manner, it is of little aid to the station operator. Of concern to all CBers is the problem of lightning protection, which is discussed first.

Lightning Protection

Over 400 people are killed each year by lightning according to the Census Bureau. Don't let your CB station make you a statistic in this toll! Lightning is the most lethal of electric discharges and is entirely unpredictable.

Lightning is a tremendous electrical spark between clouds, or between clouds and the earth, releasing millions of volts at tremendous currents in a fraction of a second. The U.S. Weather Bureau indicates that Florida and the Gulf Coast have the greatest number of lightning storms per year, followed by the central states. The West coast has the smallest number of lightning storms per year.

Regardless of where you live, don't invite lightning to strike your antenna installation. A few simple precautions may save your equipment, and even your life!

In rural areas, barns and houses have been protected from lightning for many years by "lightning rods" which seem to provide a "static electricity drain" so that electricity in the air may pass safely to

Fig. 1 LIGHTNING ARRESTOR FOR COAXIAL LINE is important antenna component for all CB stations. This inexpensive unit has precision, built-in spark gap that drains accumulated electrical charge in antenna system to ground.

ground. This makes the immediate area around the rod less conductive, thereby diverting a potential lightning stroke to an area of greater conductivity. To accomplish the same result, your tower and CB antenna should be grounded directly, or fitted with some sort of an air-gap lightning arrestor. This device has a very small gap built into it between the center conductor and the shield of the coaxial line and, when electrical charges build up in the atmosphere around the antenna, small sparks jump the gap and drain the accumulated charge to ground (Figure 1). This type of arrestor is especially important for use with a ground plane antenna whose active element projects high in the air above the ground plane elements and is thus exposed to a direct hit from lightning.

Grounding Your Coaxial Line

The coaxial cable running from your CB station to your antenna should be grounded at the station end by means of a heavy strap attached to the outer, braided jacket of the line and running to one or more ground rods driven six or more feet into the soil. Rugged, aluminum grounding wire may be used, as discussed in the next section. Finally, when a lightning storm approaches the station, it is wise to disconnect the transmission line from your equipment and to remove the line from the vicinity of your station.

If you are unlucky enough to sustain a direct lightning hit on your antenna no devices or arrestors will save the installation. The antenna and feedline will be heavily damaged or vaporized and the equipment ruined. You don't want to be in the vicinity when that happens!

Play safe! Use a lightning arrestor in your coaxial line to bleed small static charges to ground. Use a good external ground on your coaxial line. Finally, when a lightning storm approaches, disconnect the line from your equipment and, if possible, remove it from your house. Have respect for lightning and it may have respect for you!

The Ground Connection

In the early days of radio when single wire antennas were used, great attention was paid to the ground connection. Today, from an operating standpoint, the ground connection is less important when coaxial transmission lines are employed, serving principally as a conducting path between the electrical circuit and the earth, or some conducting body acting in place of the earth. This connection insures that zero voltage exists between the equipment and ground. This is very important to the CB operator as a safety measure since a ground connection limits the voltage on the equipment cabinet, microphone and antenna line to a safe value, thus reducing shock hazard. In addition, as previously discussed, the ground connection provides an easy path to the earth for static electricity and for electric currents resulting from certain equipment faults and short circuits.

The *National Electric Code* (NFPA #70) adopted by the *National Protection Association* (section 810-26) recommends that the grounding conductor for transmitting and receiving equipment should be connected to a metallic underground water piping system (the cold water pipe) or a metal ground rod driven at least six feet into the earth. The ground wire should be #4 gage copper or aluminum fastened securely to the equipment at one end and the ground rod or piping system at the other. The ground lead should be as short as possible.

It should not be assumed that your metal tower or mast is grounded if it is resting on the ground or on a concrete pad. A long ground rod should be driven into the soil and the tower or mast connected to the rod with a heavy ground wire, or strap.

The Counterpoise, or "Radio Ground"

The direct ground connection is a safety measure and should be included in all CB installations. A second artificial "radio ground", called a *counterpoise,* may prove beneficial in cases where equipment feedback (instability) or television interference is noted. In some cases,

Fig. 2 COUNTERPOISE GROUND WIRE can reduce transmitter instability and unwanted r-f on microphone cable. The insulated wire is attached to ground (chassis) of CB set and is stretched out along floor or behind operating position. Wire acts as a single radial such as is used on ground plane antenna.

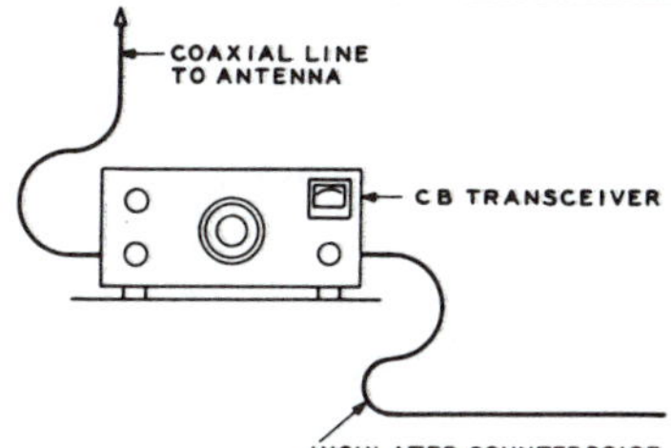

radio energy from the antenna circuit may find its way into the microphone circuit or other wiring and can cause feedback which results in erratic operation or rough modulation. A counterpoise ground often eliminates this problem. The counterpoise consists simply of a quarter-wavelength of insulated wire with one end attached to the ground post (chassis) of the CB equipment. The opposite end of the wire is left free. For the CB channels, the counterpoise wire can be about 8'6" long. The wire placement is not critical and it may be stretched out on the floor behind the operating desk (Figure 2). The wire acts much in the same manner as a single radial on a ground plane antenna to establish an electrical ground point at the equipment end of the counterpoise. Puzzling examples of transmitter instability and a "hot" microphone have been cured in a moment by this simple counterpoise which does not detract from operation of the equipment in any way.

A Simple Dummy Load

A *dummy load* is the electrical equivalent of an antenna in that it draws r-f power from the transmitter. Unlike the antenna, the dummy load dissipates the power as heat instead of radiating it into the air. The best and simplest dummy load is a resistor that duplicates the electrical load the antenna presents to the transmitter. Shown in Fig. 3 is a simple dummy load you can build. It is made of four series-parallel connected #46 flashlight bulbs which require about 6 watts of radio energy to light fully. The output of the average CB set is about 3.5 watts and when fully modulated by the operator's voice, the bulbs go from low to full brilliance.

The bulbs are wired as shown in the illustration and two pieces of short wire connect the bulbs to a coaxial plug that matches the receptacle on your equipment. Keep all leads short and direct.

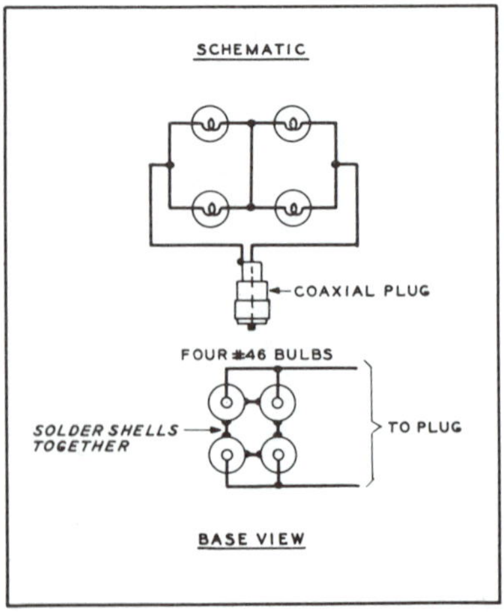

Fig. 3 DUMMY LOAD is made up of four flashlight bulbs connected in series-parallel (#46, 6.3 volt, 0.25 ampere). Shells of bulbs are soldered together and center contacts wired as shown. Bulb assembly is mounted on back of coaxial plug which matches antenna receptacle on your CB set. Dummy load is used in place of real antenna to test equipment.

The dummy load is connected in place of the antenna and the transmitter tuned and adjusted in the usual manner. Modulation may be checked by watching the increase in brilliance when talking directly into the microphone. When the dummy antenna is used, the signal from the transmitter is restricted to the immediate vicinity, but will still travel a few hundred feet from the station and may be heard weakly a few houses away.

Build A Ferrite Balun Transformer for CB

A *balancing transformer* (balun) is a transformer for converting a balanced electrical system to an unbalanced system, or vice versa. Balun transformers come in all sizes from the midget "ladder transformer" used with television antennas to giant multi-kilowatt units used in broadcasting stations.

Baluns are extremely useful in CB antenna systems when it is desired to feed a balanced antenna from an unbalanced, coaxial line. A balanced antenna is one that has two feed point terminals, neither one of which is grounded. The dipole, Yagi and Quad antennas are examples of balanced antennas with which a balun should be used. Coaxial line, on the other hand, is an unbalanced system since one terminal of the line (the outer conductor) is at ground potential.

Connecting coaxial line directly to a balanced antenna is often done, and in many cases no great inefficiency results. In the case of high gain beam antennas, however, the use of a balun is recommended. If the balun is not used, the SWR on the transmission line may rise, the SWR

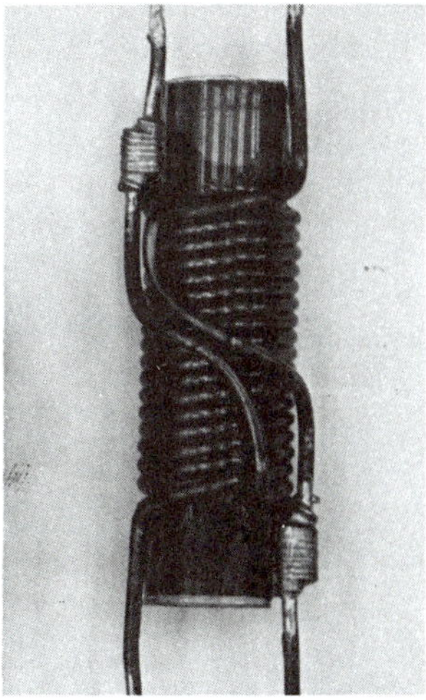

Fig. 4 FERRITE CORE BALUN TRANSFORMER is used to convert a balanced antenna system for use with an unbalanced (coaxial) feedline. This inexpensive balun consists of three windings of #14 enamel copper wire wound on a ferrite slug. Windings are interconnected as shown in Figure 5. See Chapter 14 for information on purchase of slug, or consult your nearby electronics distributor.

reading may be inaccurate, or the polar pattern or front-to-back ratio of the beam may be seriously affected. Feedline interaction of this sort is termed *antenna effect* and is most noticeable on arrays having high gain and good front-to-back ratio.

Placing a balun transformer between a coaxial feedline and a balanced antenna (at the antenna) permits the coaxial line to perform its proper duty, that of transporting radio energy from one place to another and prevents it from becoming part of the antenna itself.

A Practical Balun

A simple balun transformer that does the job is shown in Figure 4. It will match a balanced antenna to a 50 or 70 ohm coaxial line without disrupting the existing SWR on the transmission line. Power handling capacity of the balun is about 150 watts at CB frequencies. The balun consists of three windings placed side by side on a short, ferrite core. This type of winding is called a *trifilar* winding.

The balun shown in the photograph consists of three windings, each 6 turns of #14 enamel wire wound on a Q-1 ferrite slug. The ferrite slug is 1/2-inch in diameter and about 3¼ inches long. It is a section of an *Indiana General CF-503* slug which is 7½ inches long. The ferrite is broken to length by nicking it with a file around the circumference at the desired length and breaking it with a sharp blow. Here is how you make the special windings:

The simplest way to do the job is to make the three windings as one. Cut three pieces of #14 enamel copper wire, each about 3 feet long.

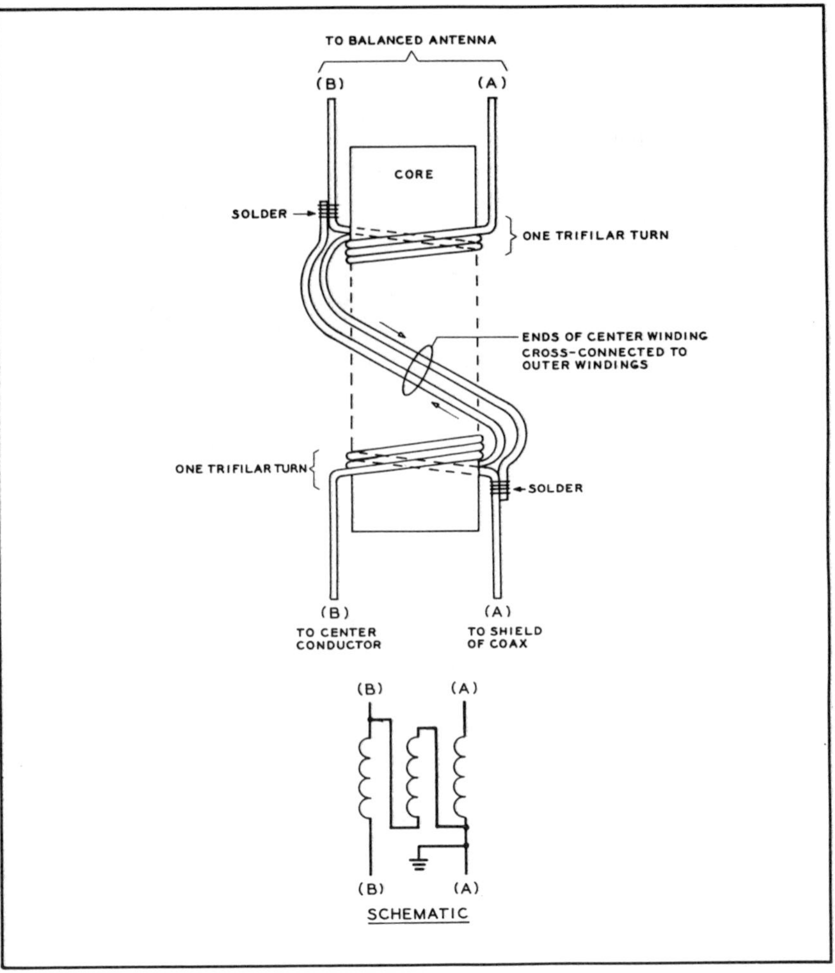

Fig. 5 WINDING TECHNIQUE FOR FERRITE CORE BALUN TRANSFORMER. Three parallel windings are placed on the ferrite slug. The windings are wound on as one wire. The center winding is cross-connected to the opposite ends of the outer windings to equalize current flow. Tape or cement ends of the windings. Balun should be mounted at the antenna terminals in a waterproof container.

Place one end of the wires in a vise and smooth out the sections until they are parallel. Grasp the free ends in your hand and wind them side-by-side on the ferrite core as if they were one wire. Wind the wires under tension and they will remain in position. When you are finished, dress the leads away from the core as shown in the photograph.

You'll note that the center winding is cross-connected to the oppos-

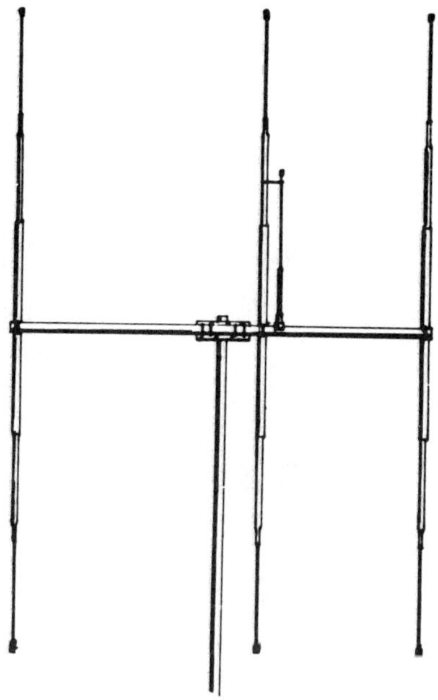

Fig. 6 GAMMA MATCH transformer is often used with Yagi beams. Gamma transformer rod is mounted parallel to driven element, a few inches away from it. Far end of rod is attached to antenna with a movable jumper. The near end of rod is connected to the center conductor of the coaxial feedline through a variable capacitor. Capacitor is often placed within the Gamma rod for protection from the weather, as in this typical arrangement. See Fig. 7 for details.

ite ends of the two outer windings (Figure 5). Either end of the balun may be taken as input or output, but the common connection between the inner and outer winding at the end you choose as the input end must be taken as a ground point and attached to the outer shield of the coaxial line.

Cut the ends of the inner winding to length, scrape off all the enamel and make the solder connections to the outer windings, as shown in the photograph. A drop or two of coil dope, Krylon or nail polish may be put on the ends of the windings to anchor them to the core. Do not coat the windings themselves with any material as this tends to upset the balance of the transformer.

You should protect the balun from the ravages of the weather. This can be done by placing it within a plastic bottle and coating the entrance holes with *General Electric RTV-102* or other waterproofing compound.

The Gamma Matching System

A transmission line terminated in other than its characteristic impedance reflects a portion of the transmitted power back to the input end of the line. In most cases, simple antennas do not present the exact

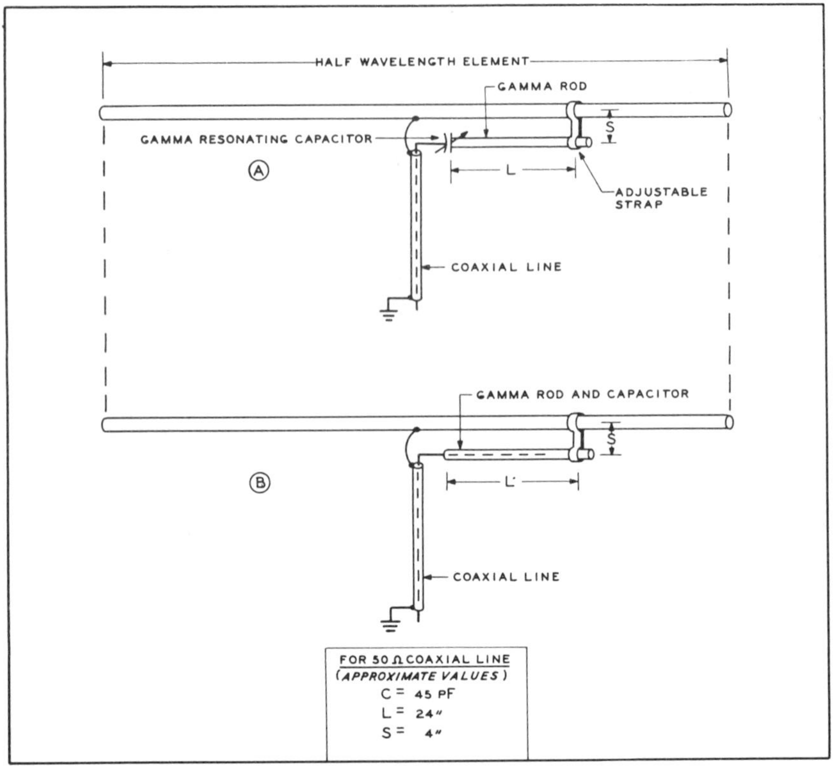

Fig. 7 GAMMA MATCHING SYSTEMS. Basic Gamma matching system using a variable capacitor is shown in (A). Both terminals of capacitor are insulated from driven element and boom. Diameter of Gamma rod should be about ½ to ¼ that of driven element. Section of coaxial line may be modified to serve as a combination rod and capacitor (B). Length of center wire provides capacitance.

value of termination to a coaxial line that provides a low value of reflected power (SWR). Thus, SWR values of 1.5 to over 2 are commonly noted on many antennas used in CB service. Sometimes the mismatch is ignored by the antenna designer. In other cases, a matching transformer known as a *Gamma match* is used to transform the radiation resistance of the antenna to a value that more closely fits the radiation resistance of the transmission line.

A typical Gamma match system is shown in Figures 6 and 7. It consists of a metal transformer rod mounted parallel to the antenna and a few inches away from it. The far end of the rod is attached to the antenna a few feet out from the center point, and the near end of the rod is attached to the center conductor of the coaxial transmission line through a tuning capacitor. The shield of the line is connected to the center point of the antenna. The transformation ratio of the Gamma match is determined by the length of the rod, the spacing of the rod from the antenna, and the ratio of the diameter of the rod to that of the antenna element. Suitable dimensions for the CB channels are shown in the illustration.

The Gamma match is adjusted with the aid of an SWR meter placed in the transmission line. Power is applied to the antenna and the length of the Gamma rod and value of the capacitance are adjusted until a very low value of SWR is obtained. When properly adjusted, the Gamma match can provide an SWR of less than 1.2 at the resonant frequency of the antenna.

In addition to providing a variable transformation ratio, the Gamma match also acts as a form of balun, matching the unbalanced coaxial transmission line to the balanced driven element of the antenna.

Television Interference (TVI)

Your CB transmitter can, and often does, cause interference to nearby television receivers, especially in fringe areas where the television signal is weak. There are various causes of television interference, and one of the most common is the impairment of the TV picture by *harmonic signals* radiated by the CB transmitter. The harmonic signal is one that is a multiple of the transmitted frequency. For the CB transmitter operating on 27 MHz, the second harmonic frequency is 27 x 2 = 54 MHz, the third harmonic frequency is 27 x 3 = 81 MHz, and the fourth harmonic is 27 x 4 = 108 MHz. A small amount of r-f power may be generated by even a well-designed transmitter at any one of the harmonic frequencies and it does not take much harmonic power to override a TV picture, if the harmonic energy happens to fall on the same frequency as the TV picture transmission. In particular, picture transmission on TV channel 2 centers about 55.25 MHz, which is uncomfortably close to the second harmonic of the CB frequencies. The harmonic signal of the CB transmitter will cause a succession of light and dark bands across the TV picture often referred to as a *cross-hatch* or *herring-bone* pattern superimposed on the picture. In some cases, in addition to this interference pattern, the CB transmitter in close proximity to the TV receiver may cause *sound bars* on the received picture which result from voice modulation of the transmitter. These are horizontal bars that vary with voice modulation.

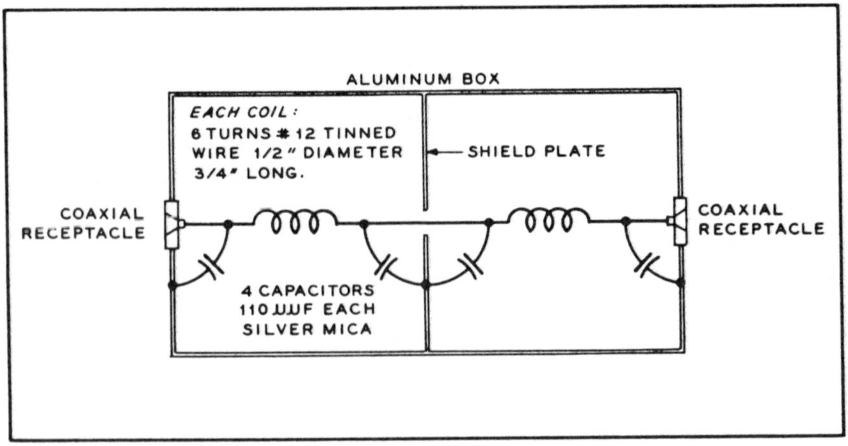

Fig. 8 SIMPLE TVI FILTER FOR YOUR CB TRANSMITTER. This filter prevents the TVI-producing harmonics in your transmitter from reaching the antenna. The filter is built in an aluminum "mini-box" measuring 4" x 2" x 2-3/4". A BUD type CU-2115A will do the job. Place a shield plate across the inside of the box as shown, with a 1/2" hole in the center. Bolt the plate in position with 4-40 screws, lockwashers and nuts. The coils are self-supporting and are soldered directly to the center terminals of the coaxial receptacles. Wrap insulation around the wire that passes through the shield hole to prevent a short circuit. Bolt the box together securely with self-tapping screws.

It is impractical to build a CB transmitter that does not have *any* harmonics, but some of the units are much better than others at suppressing or eliminating the harmonic currents within the transmitter circuitry. Efficient harmonic suppression may be achieved by the addition of a simple *harmonic filter* placed permanently in the coaxial transmission line between the CB equipment and the antenna. This filter prevents the harmonics in the equipment from being radiated by the antenna. Experience shows that the use of a good harmonic filter will, in the majority of instances, suppress or eliminate television interference caused by CB equipment of modern design. Well designed *low-pass filters* suitable for CB and radio amateur use are available from a number of manufacturers, or you can build an effective and inexpensive filter. (This unit is called a *low-pass* filter because it passes the lower CB frequencies but stops the higher harmonic frequencies).

Building Your TVI Filter

Shown in Figure 8 is a pictorial representation of a simple and effective low-pass TVI filter suitable for use with any CB equipment. It

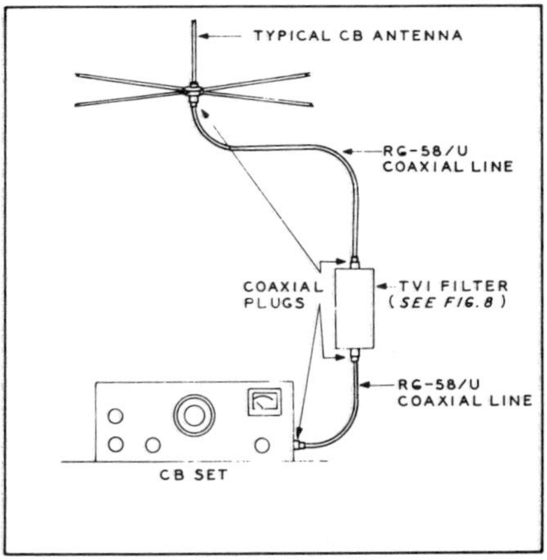

Fig. 9 TVI FILTER is installed in coaxial line between CB set and the antenna. It is usually placed close to the transmitter. Coaxial plugs (properly installed) should be used at all joints in the coaxial line. If an SWR meter is used, it should be placed in the line between the filter and the CB set.

is designed to suppress harmonic energy generated by your transmitter that might reach your CB antenna and be radiated to a nearby television receiver. Specifically, the filter reduces the harmonic energy output of the transmitter by a factor of 1000. The filter is built in a small aluminum box and fitted with two coaxial receptacles which match the coaxial fittings on your transmission line. Either end of the filter may be taken as input or output, as the unit is symmetrical.

The filter is composed of two circuits, each in a separate compartment, connected together by a single wire which passes through an insulated hole in the shield. The four capacitors shown are readily available units, and the small coils may be hand-wound out of enamel covered copper wire. The illustration suggests placement of parts within the box, and it is suggested that the general layout be followed closely. All leads and connections between the components should be as short as practical, and properly soldered. The lid, or cover, of the box should be held firmly in place by means of sheet metal screws to prevent harmonic signals from leaking out along the seams.

Using the TVI Filter

The filter is placed in the 50 ohm coaxial line running from your CB station to your antenna. Normally, the filter is located within a few feet of the equipment. Use coaxial connectors at all joints in the line, as shown in Figure 9 so that r-f energy cannot "leak" around the filter. If you want to include an SWR meter in your antenna circuit, place it between the filter and your equipment rather than after the filter.

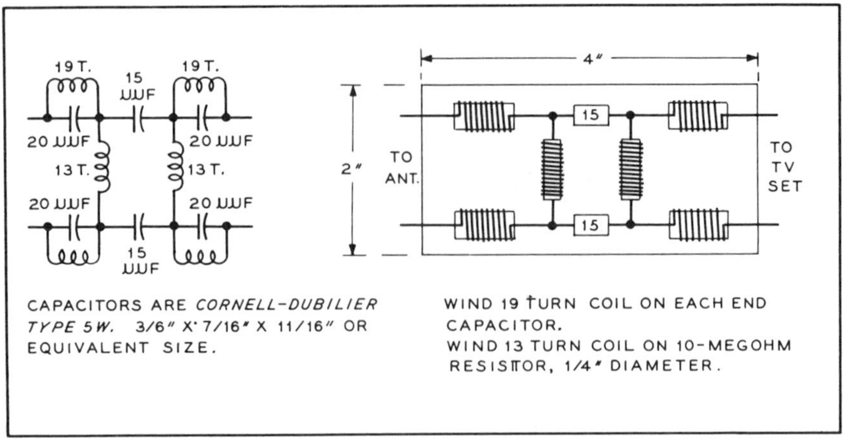

Fig. 10 HOMEBUILT TVI FILTER FOR YOUR TV SET. Harmonics from your CB transmitter can be stopped before they reach your TV set by adding this simple filter in series with the "ribbon line" from the TV antenna to your receiver. The filter is built upon a small board made of thin plastic or phenolic material. The end coils are wound directly over the body of the capacitors, with the turns spaced the diameter of the wire. The center coils are close wound on dowel rod or resistors used as coil forms. Mount the filter close to your TV receiver.

A Filter for Your TV Set

Even though you use a low-pass TVI filter on your transmitter it is possible for the signal from the CB transmitter to overload a nearby television receiver. This comes about because the circuits of the television set are not selective enough to reject a strong, nearby signal operating close to the TV channels. Since the TV channels start at 54 MHz and extend upwards in frequency, it may seem absurd to think that a signal on 27 MHz is "close" to the TV channels. However, your CB signal may be as much as 10,000 times as strong as the signal from the TV station, and it is asking a lot of some inexpensive TV sets to reject such a strong signal, regardless of its frequency.

Shown in Figure 10 is a simple filter that is placed in series with the 300 ohm ribbon lead from the antenna to your television set, at the terminals of the TV set. The filter rejects unwanted signals below TV channel 2 and passes all signals above (and including) channel 2.

The filter is small and may be made self-supporting, using the leads of the various components to support the assembly. It is composed of six capacitors and six small coils, interconnected with very short leads. The parts are assembled exactly as shown in the right-hand illustration of Figure 10. A phenolic or plastic board 4'' x 2'' is cut, and the completed filter assembly is fastened to the board by drops of epoxy cement. The four end circuits are made by winding the coils over the bodies of the capacitors. Nineteen turns of #24 enameled wire are wound on each capacitor, the turns spaced approximately the wire diameter. The ends of the windings are cleaned and the leads soldered to the capacitor wires. Each center coil is made up of 13 turns of the same size wire wound around a 10 megohm, 1 watt resistor used as a coil form. A 1/4-inch diameter length of wood dowel rod may be substituted for the resistor form, if desired. The turns are spaced the diameter of the wire. The dowel may be drilled with a small drill or needle and the end turn of each coil passed through the hole to lock it in place.

When the filter is completed and fastened to the phenolic board, the 300 ohm TV ribbon line from the TV antenna is soldered to one set of terminals of the filter. The other set of terminals are attached to the antenna connections of the TV set. If possible, the filter should be mounted inside the set, positioned as close to the TV tuner as you can possibly get so that the leads from the filter to the TV set are short.

Measuring R-F Loss in Coaxial Line

The efficiency of your coaxial line may be determined by measuring the line loss at the CB frequencies. You can do this by merely short circuiting the far end of your coaxial line and measuring the standing wave ratio with your SWR meter. If there is no line loss whatsoever, the SWR reading will be infinite (full scale), indicating that the reflected wave is equal in amplitude (size) to the incident wave. In a real-life situation, of course, this is not the case, and the SWR reading under the test condition will be less than infinite, due to line loss.

In order to make this measurement, the antenna termination is removed from the far end of the transmission line and the outer shield is firmly shorted to the inner conductor of the line. A small amount of power is applied to the line through the SWR meter. The meter is adjusted for full scale reading on the "forward" position, and the meter switch is then thrown to the "reverse" position. The line loss may then be computed from the reverse reading and the chart in Figure 11. If, for example, the SWR turns out to be 4.5, the cable loss (attenuation) is 2 decibels. This means that your coaxial line is about 63 percent efficient, and that 37 percent of your transmitter output power is being lost in the line. If the SWR reading, on the other hand, is 9; then your line loss is only 1 decibel and your line is about 80 percent efficient.

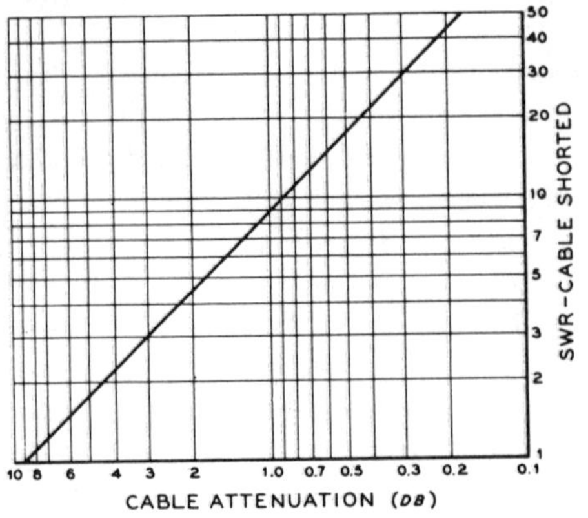

Fig. 11 COAXIAL LINE LOSS may be determined by simple test. End of line is shorted and SWR is measured under shorted condition. For example, if SWR is 10, line loss is 0.9 decibel.

A word of warning: this test should not be conducted with a CB transmitter having a solid state (transistor) output stage because it is possible to damage the transistor under these conditions. Equipment using a vacuum tube in the transmitter output stage can withstand this unusual form of operation with less chance of damage, particularly if the duration of the test is only 30 seconds or less.

Helpful Pointers on Soldering

A large percentage of antenna (and equipment) trouble arises because of poorly soldered joints. Working with CB antennas can be great fun, and it's worth your time and effort to make sure that the job you do is done properly. A poorly installed coaxial plug on a feedline can short out your whole antenna system and possibly cause expensive damage to your CB equipment. Improperly soldered connectors in your coaxial line or in the antenna itself can bring deafening static, or perhaps no signals at all! All of these problems can be avoided if you know how to solder correctly and do a ship-shape job on each individual connection.

Many CB service shops report that the majority of troubles with CB antennas comes about from improperly soldered coaxial connectors. Don't let this commonplace problem be your downfall!

Soldering is not merely glueing or sticking two metals together. When correctly done, it unites the metals so that they can be considered to be one piece as far as the electrical current is concerned.

The Soldering Gun

For general work on coaxial lines and connectors, a dual heat (145-210 watt), medium size soldering gun is recommended. Don't use a low wattage "pencil iron" for antenna work, as this tool doesn't have the wattage to properly heat the coaxial plugs and fittings which have a great deal of thermal mass. A large, high wattage soldering gun, on the other hand, can overheat coaxial cable and melt the inner insulation. A gun in the 140 to 250 watt range seems to be best for all-around antenna work.

If the soldering gun is new, read the instructions and prepare the tip for use. If the gun is old and the tip is dirty or has had a lot of use, prepare it by filing it with a fine file until the working surface is shiny. Then heat the gun and coat the tip with rosin core solder. *Never* use acid core solder as the acid can quickly ruin the gun tip and the equipment you are working on.

The Joint

When the gun is ready to be used, the joint should be prepared for soldering. The two metals to be joined should be completely clean. Remove all dirt, enamel, scale or oxidation by sanding or scraping down to the clean metal. Insulation may be removed from wires with the blade of a pocket knife. (Make sure you do not nick the wire).

The first step is to tin the metal surfaces. Hold the hot tip of the gun against the metal until the solder melts and flows onto the clean material. When properly tinned, the metal should be covered with a thin coat of solder, and the excess solder will flow onto the tip of the gun. Apply the solder to the joint, don't try to bring it to the joint on the tip of the gun.

When you are working with two wires, or wire leads, you should make a good mechanical joint between the parts being soldered. This doesn't mean that you should wrap the wires tightly around each other, or wrap a wire lead many times around a soldering terminal or lug. If you do this, you will have a hard time removing it at a later date if you wish to make a change in the circuit. Rather, bend the lead around the terminal once--just enough to hold the two pieces together, then solder the pieces. Antenna joints, on the other hand, should be well wrapped before soldering to overcome the stresses and strains in the wire under the continuous action of the wind.

The final step is to apply the soldering gun to the connection. Hold the tinned surface of the tip against the joint and let it come up to temperature. When it reaches the proper level, solder applied to the joint will flow quickly and smoothly over the joint. Never try to solder by applying solder to the iron and then to the joint, or by applying a lump of solder to the joint and then pressing it down with the gun.

A little experience will teach you just how hot the joint should be. If it is not hot enough, the solder will have a grainy appearance and the joint may snap apart. This is called a *cold solder joint.* If the joint temperature is excessive, you may damage the components you are working with, or melt the inner insulation of the coaxial cable. Most beginners err on the side of cold joints, as they are impatient and do not take the time to let the joint come up to the proper temperature.

Soldering Outdoors

Soldering connections on an antenna installation in cold weather can be an exasperating experience as the cold air robs the soldering gun of heat at a very fast rate. To overcome this, the easiest way to bring the antenna joint to the proper temperature is to have an assistant with a second soldering gun at hand. The application of the second gun to the joint for a few minutes will bring the connection up to soldering temperature. A blanket shielding the joint from the wind will help, too.

A Word to the Wise

Watch out when you solder coaxial braid! One or two unsoldered hair wires of the braid can easily short out a coaxial plug, thus shorting out the whole antenna system. Care taken in soldering plugs can pay big dividends in the long run.

Finally, remember that it is impossible to solder aluminum. In spite of so-called aluminum soldering compounds being sold on the market, no satisfactory technique has been developed to solder this metal because of its ability to oxidize rapidly in the air. The only reliable aluminum joint is one that is mechanical, the pieces of aluminum being held together by pressure.

Before the aluminum joint is made, the metal should be lightly coated with a special anti-oxidizing compound (*Penetrox,* manufactured by Burndy Co., Norwalk, Conn). Small tubes of *Penetrox* may be obtained from many antenna manufacturers. This paste forms an air-tight seal, preventing corrosion. The compound is a good electrical conductor and will insure a trouble-free joint.

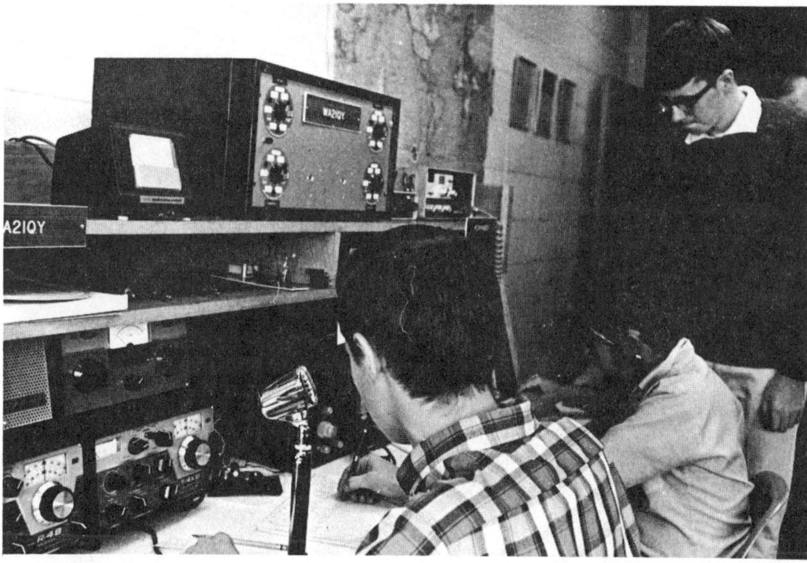

DURLAND EXPLORER SCOUT CENTER, Rye, NY has code classes and technical assistance for beginners. W2NVB and KCK 4920 are operated by Scouts. Check your Scout Center to see if Novice amateur classes are held. Many radio clubs also conduct beginner classes and some amateur stations transmit code practice sessions. It is easy to get your amateur license if you really want it!

Safety First!

Each year a number of CBers are killed working on their antennas. Some of them lose their grasp and fall to the ground. Others are electrocuted when part of the tower, antenna or guy wires accidentally touches a high tension line. Don't become a statistic. Plan ahead! Wear a safety belt if you climb a tower. Make sure the tower is well guyed before you climb it. If your antenna is near a power line, make sure that it cannot hit the line if it falls. Never attach an antenna or guy wire to a utility pole!

If your antenna touches a power line, don't touch the antenna, tower or feedline. You could be electrocuted! Call the power company at once. Finally, learn first-aid and artificial respiration. Your quick action may save the life of a friend. Think before you act. The life you save may be your own!

Glossary of Antenna Terms

(Say What You Mean--Mean What You Say)

Many CBers are bewildered by the avalanche of new technical terms that are thrust at them in this new era of communications. This glossary section is an analysis of some of the more common words, terms and phrases used in the world of CB, as determined by common usage. Missing from the list are all of the "nonsense phrases" used by some manufacturers to extoll the virtues of their products. For those serious CBers desiring an extensive list of over 16,500 electronic terms, the "Modern Dictionary of Electronics", edited by Rudolph F. Graf and published by Howard W. Sams & Co., Indianapolis, Indiana is suggested.

* * *

Active antenna element - the element which receives power directly from the transmitter; also, "driven element."

Alternating current (AC) - a flow of electricity which reverses itself many times per second. House current has a frequency of 60 Hertz (Hz), or 60 cycles per second.

Amateur Radio Operator - an individual licensed by the FCC to build, install, operate and experiment with radio equipment both as a hobby and as a public service; having no pecuniary interest in his work. Amateurs also are called "radio hams."

American Radio Relay League - an organization which furthers the interests of amateur radio; located in Newington, Conn.

Antenna - a conductor (usually wire or metal tubing) suspended in space to radiate and receive electromagnetic waves.

Antenna array - a system of antenna elements, or antennas, which produces high gain in transmitted and received signals.

Antenna gain - the effectiveness of a directional antenna in a particular direction, compared against a standard antenna such as a dipole.

Antenna mount - the mechanical device to which the base of an antenna is attached to secure it to a vehicle or other support.

Antenna pattern - see "Polar Plot."

Antenna tuner - a circuit used to "match" a transmitter or receiver to an antenna system (see "Matching transformer").

Attenuate - to weaken, reduce in power or strength.

Balun - specially wound coils which transform an unbalanced r-f system into a balanced system, or vice versa; used in feeding transmitting antennas.

Bandwidth - the range of frequencies over which an electrical circuit or device will respond efficiently; also, the upper and lower limits of a band of frequencies.

Base-loaded antenna - an antenna which has a loading coil at its base to increase its electrical length.

Beam antenna - an antenna with several elements that concentrates its radiation into a narrow beam in a definite direction.

Bidirectional antenna - an antenna which radiates two main lobes, usually in a figure-8 pattern.

Bonding - placing metal objects at the same electrical potential by connecting them together with heavy wire or braid. Such objects are usually at ground potential.

Boom - the member of a beam antenna which supports the active and passive elements.

Capacitive reactance - the resistance offered by a capacitor to the passage of an alternating current. Reactance is not a pure resistance even though it is measured in ohms.

Capacitor - (condenser): a circuit component consisting of two or more plates separated by a dielectric (insulator).

CB (Citizens Band) - the no-examination radio service established by the FCC to permit low-cost, short distance, business-type radio communication.

CB channel - each specific frequency set aside by law for CB communications.

Center-loaded antenna - an antenna with its loading coil at its center to increase its electrical length.

Choke coil (r-f) - an inductance (coil) which stops the passage of r-f.

Circuit - an interconnection of electrical components to perform a specific electrical function (detection, amplification, etc.)

Coaxial antenna - an antenna which uses coaxial line for all, or part, of its length.

Cobra antenna - a half-wave vertical dipole antenna at the base of which is a ferrite choke acting as a decoupling coil.

Coil - a number of turns of wire wound around an insulating form (or self-supporting) to form an inductance.

Cold flow - the gradual movement of a wire through an insulating material caused by bending the wire too sharply.

Cold solder joint (cold joint) - a poorly made solder connection in which solid metallic contact has not been achieved; can cause non-operation of equipment.

Conductor - a metal containing numerous free electrons which conducts electricity.

Corrosion - the decomposition of metals which takes place in salt air.

Counterpoise ground - an artificial "radio ground" composed of one or more wires either near the transmitter or strung out under the antenna in the form of a ground plane.

CPS - cycles per second; now called Hertz (Hz). One Hertz is equal to one CPS.

Cross-polarization - the existing condition when one antenna is vertical and another antenna is horizontal.

Cubical Quad - a type of beam antenna consisting of two or more loops of wire spaced approximately a quarter-wavelength apart. One loop is driven by r-f power from the transmitter, the other may be a reflector or director.

Cycle - an alternating current is said to complete one cycle when the movement from zero to maximum in one direction, and from zero to maximum in the reverse direction, is completed.

Decibel (db) - a unit of measurement which expresses the ratio between two power levels; a change of 1 db. is just detectable as a change in loudness under ideal conditions.

Decoupling coil - an inductance which electrically isolates one electrical circuit from another.

Demi-Quad antenna - a Quad-type antenna with only one wire loop.

Dipole - a half-wave antenna; in the 27 MHz CB band, a dipole is about 18 feet long.

Director - the element of a beam antenna placed in front of the driven element which reradiates a portion of the transmitted energy in the desired direction.

Double cross array - see "Cross Polarization".

Driven element - an antenna element which receives power directly from the transmitter via a transmission line.

Dummy load - a resistor which duplicates the electrical characteristics of the antenna without radiating a signal.

Electrolytic corrosion - the deterioration of metals in sea water resulting from an electric current applied to the metals.

Electromagnetic spectrum - the electric waves found in nature, ranging from very long waves at very low frequencies to extremely short waves at very high frequencies.

Electromagnetic wave - radiant energy consisting of electric and magnetic force fields radiated into space.

F-1, F-2 layers - the layers of the ionosphere which reflect 27 MHz CB signals back to earth.

Federal Communications Commission - the federal agency which regulates U.S. communications, including CB and amateur radio.

Feed - (in radio): to supply r-f power to an antenna.

Feed impedance - see "Radiation resistance".

Feedline - see "Transmission line."

Ferrite core - a circular ring of powdered iron placed inside a coil to raise its inductance.

Field pattern - see "Polar Plot."

Figure-8 pattern - the bidirectional radiation pattern of a half-wave dipole.

Five-eighths wavelength antenna - an antenna which is 5/8ths of one wavelength long at its resonant frequency.

Free space wave - a radio wave which travels from the antenna outward into space.

Frequency - the number of times a second that an alternating wave completes one cycle (one Hertz).

Front-to-back ratio - the ratio of power radiated in the forward direction by a beam antenna to the power radiated in the back direction.

Gain (forward gain) - the maximum signal power amplification measured in decibels provided by a particular antenna, compared to a reference antenna.

Galvanic corrosion - the deterioration which occurs when two dissimilar metals are placed close together in salt water.

Gamma match - a type of matching transformer used to change the radiation resistance of an antenna to that value required by the transmission line; employs a metal rod and variable capacitor located next to the driven antenna element.

General coverage antenna - see "Omnidirectional antenna."

Grid dip meter - an instrument which indicates by the deflection of its meter when it is tuned to the frequency of a circuit near which it is held.

Ground - (in radio): (1) wires and components connected electrically to the chassis of the equipment form a "chassis ground"; (2) equipment connected by heavy wire to a pipe driven into moist earth is "earth grounded."

Ground plane antenna - a vertical quarter wavelength antenna mounted above several horizontal rods or wires (radials) that form an artificial ground.

Groundplate - (marine): a copper plate fastened to the hull of a boat underwater to provide a ground for the boat's electrical equipment.

Ground wave - that portion of a radio wave which follows the curvature of the earth.

Half-wave antenna - an antenna with a physical length approximately equal to one half-wavelength at its resonant frequency (about 18' at 27 MHz).

Harmonic - an integral multiple of the fundamental frequency, i.e., the fundamental frequency multiplied by 2, 3, 4, etc.

Hertz - (Hz) one cycle per second.

High pass filter - a circuit which permits passage of frequencies above a chosen frequency and blocks frequencies below the chosen frequency.

Horizontal polarization - a radio wave is horizontally polarized when its electric field is parallel to the earth's surface.

Impedance - the opposition a circuit offers to the flow of alternating current. It is measured in ohms even though it is not a true resistance.

Impedance matching - the process of adjusting an impedance so that maximum power flows through the circuit and (in an antenna circuit) minimum SWR exists.

Incident wave - the r-f power flowing on a transmission line towards the antenna.

Inductive reactance - the resistance offered by an inductor to the passage of an alternating current.

Input impedance - see "Radiation resistance."

Ionosphere - the electrically charged layers of air existing 60-250 miles above the earth which reflect high frequency radio waves back to earth.

Isolated stay antenna - a stay or shroud on a sailboat in which insulators have been placed to permit its use as an antenna.

Isotropic antenna - an imaginary antenna considered to be so small that it radiates energy equally well in all directions.

Kilohertz (kHz) - 1,000 Hertz (cycles per second).

Lightning arrestor - a device placed in the transmission line containing spark gaps which allows lightning currents to flow to earth.

Load - a device (such as an antenna) connected to the output of a transmitter to absorb r-f power.

Loaded whip antenna - an antenna which uses an inductance (coil) to provide additional electrical length to the antenna.

Loading coil - an inductance at the base or center of an antenna which makes the electrical length of the antenna greater than the physical length.

Lobe - an area of greater signal strength in the pattern of a beam antenna.

Low pass filter - a circuit which permits passage of frequencies below a chosen frequency and blocks frequencies above the chosen frequency.

Marine ground - see "Ground plate."

Matching transformer - an electrical device which transforms the impedance of one circuit to that of a second circuit.

Megahertz (MHz) - 1,000,000 Hertz (cycles per second).

Mini-whip antenna - a very short, loaded vertical antenna.

Monster Quad - a Quad antenna with four or more elements.

Node - the point on an antenna system at which some characteristic has zero amplitude.

Null - a point of zero voltage or current; the minimum reading on a meter.

Ohm's Law - the law expressing the ratio of the voltage to the current as measured across a known resistance or impedance. $R = E/I$, or $E = IR$, or $I = E/R$.

Omnidirectional antenna - an antenna which radiates and receives energy equally well in all directions (general coverage, or non-directional antenna).

Parasitic element - an antenna element which does not receive power directly from the transmitter.

Phase - the relationship, at a given point in an a-c circuit, between voltage and current.

Phased antenna array - see "Stacked antennas."

Pi-network - a type of transmitter output circuit capable of matching a range of antenna impedances.

Polar plot - a graphic representation of the magnitude of radiation from an antenna expressed in terms of the distance and direction from the antenna.

Polarization - the position and direction of the electric field of a radio wave with respect to the earth's surface.

Power gain - the increase in radiated power provided by one antenna compared to that of a standard reference antenna, such as a dipole.

Pretuned - a circuit or device (such as an antenna) which is adjusted by the manufacturer to a desired frequency.

Prune - (radio): to shorten the length of an antenna to make it resonant.

Q-factor (Q) - the figure of merit which denotes electrical efficiency. High-Q denotes high efficiency, low-Q denotes low efficiency.

Quad - see "Cubical quad."

Quarter-wave antenna - an antenna with a physical length approximately equal to one-quarter wavelength at its resonant frequency.

Radial - the horizontal wires or rods which fan out from the base of a vertical antenna to form an artificial "radio ground."

Radiation pattern - see "Polar plot."

Radiation resistance - the ratio between the voltage field and the current flowing at any given point along an antenna, expressed in ohms.

Reactance - see "Impedance."

Reflected power - r-f power reflected back down a transmission line due to an impedance mismatch between feedline and antenna.

Reflector - the element of a beam antenna placed behind the driven element which reradiates a portion of the transmitted energy forward through the driven element.

Resonance - the condition in which inductive and capacitive values in a circuit are such that maximum current flows in the circuit.

Resonant frequency - (antenna): the frequency at which the characteristics of an antenna are in electrical balance. Resonance is determined by the electrical length of the antenna.

Ribbon line - two wire antenna lead-in cable for general TV service.

Series connected - components so connected that the output lead of one is attached to the input lead of the next.

Shield - a metallic braid or metal sheet which isolates one electrical conductor or circuit from another.

Short circuit - a direct connection between two parts of an electrical circuit not normally connected together.

Skip distance (skip signals) - those portions of the earth in which a h-f signal cannot be heard because it has not yet returned to earth from the ionosphere; the signal "skips" over these areas of the earth.

Sky wave - the radio waves returned to the earth from reflection by the ionosphere.

Sporadic-E layer - one of the irregular layers of the ionosphere which reflects radio waves back to earth at intervals during the year.

Stacked antennas - two or more beam antennas placed one above another, or side by side, to obtain increased gain.

Standing waves - electric variations in voltage amplitude formed on a transmission line by energy reflected by a mismatch between load (antenna) and line.

Standing wave ratio - (also SWR, or VSWR - voltage standing wave ratio): the ratio of size (amplitude) of the outward going wave to the size of the reflected wave on a transmission line; the ratio of maximum to minimum voltage or current along the line.

Sun spot cycle - cyclic changes in radiant energy from the sun which cause changes in the ionosphere above the earth, thus affecting radio propagation.

Surge impedance - see "Radiation resistance."

SWR bridge - an instrument and indicating meter which monitor the standing wave ratio on a transmission line.

SWR curve - the SWR on a transmission line measured over a range of frequencies and plotted in the form of a curve on graph paper.

Tin - (soldering): to cover the tip of a soldering gun, or iron, or work to be soldered, with a thin coat of melted solder.

Transmission line - the wire(s) which carry r-f power from transmitter to antenna, and received signals from antenna to receiver.

Trifilar winding - a three winding coil having all windings physically in parallel and wound in the same direction.

Truth Table - the copyrighted table in this Handbook which compares the gain of ten different types of CB antennas.

Turnstile antenna - two dipole antennas at right angles to one another; the radiation pattern being nearly omnidirectional.

Unidirectional antenna - an antenna which radiates primarily in one direction.

Unity - the condition on a transmission line when no standing waves exist; the SWR is unity, or 1.

Vertical antenna - an antenna with its principal radiating element perpendicular to the earth's surface.

Vertical Polarization - a radio wave is vertically polarized when its electric field is vertical with respect to the earth's surface.

Wavelength - the distance from one crest to the next of an electromagnetic wave; the distance a radio wave travels during one Hertz (one cycle per second).

Whip antenna - a thin, vertical, self-supporting antenna.

Wing spread - (slang): the amount of horizontal space a beam antenna occupies.

Yagi beam - a widely used form of beam antenna employing parasitic elements; named after its inventor, Dr. Yagi, a Japanese scientist.

INDEX

Adapter, coaxial, 94-99
Adjustment, Mini-whip antenna, 117-120
Aluminum tubing, 181-184
Amateur radio, 214-224
Antenna
- assembly techniques, 149-194, 181-184, 210-212
- base station, 149-194
- beam, 23, 42, 51, 149-194
- Bi-square, 160-163
- coaxial, 43-45
- Cobra, 154-156
- coil loaded, 47, 113-118
- corrosion, 138, 140-142
- cross-polarized, 56, 64, 74
- definition of, 10
- design considerations, 42-53, 136, 181, 188-190
- dipole, 21, 43, 70-73, 154
- directional theory, 50
- dish, 11
- Demi-quad, 61, 164-168
- dummy, 100, 199
- electrical length, 19
- end effect, 19
- extended vertical, 47
- five-eighths wave, 47, 48, 74
- front-to-back ratio, 24
- gain, 25, 54, 67-75
- Gamma match, 203-205
- ground plane, 45, 70-74, 143, 150-154
- Ground Rules, 71
- gutter clamp, 128
- guying, 149, 151
- half-wave dipole, 49
- helix, 132
- isotropic, 24, 69
- load, 22, 35-40
- location, selection of, 26-28, 149
- loss, 116
- marine, 135-148
- matching, 22, 35-40, 91, 104, 105
- Mini-whip, 113-116, 125
- Monster Quad, 62, 80-82, 187-194, 219
- mobile installations, 111-134
- mounting techniques, 120
- Quad, 59-65, 164-172, 187-194
- radiation patterns, 50, 54, 188
- reradiation, 21
- resonance, 15-17, 21, 119
- sailboat installation, 136, 144
- soldering techniques, 210, 211
- stacked arrays, 58, 59
- trunk mount, 126
- Turnstile, 156-160
- vertical gain, 74
- whip, 47, 111-134
- Yagi, 50, 53-59, 173-180, 184-186

Arrestor, lightning, 146, 196

Balancing sleeve, 158, 176, 177
Balun transformer, 63, 193, 200-203
Base station antennas, 164-194
Beam antenna, 22, 42, 51-55, 149-194
Bi-square beam, 160-163
Boom, 56
BNC fitting, coaxial, 95
Bumper mount, antenna, 120

Capacitive reactance, 115
Center loading, antenna, 115-117
Characteristic impedance, 91
Choke, ferrite r-f, 154, 170, 193, 200-203
Coaxial adapter, 98
Coaxial cable, 87-101, 122
- contamination, 89
- cutting technique, 96, 152
- ground, 197
- impedance, 34, 91
- line check, 99
- loss, 89, 101, 209
- mismatch, 102
- resistance, 92
- waterproofing, 93

Coaxial plugs, 94-97
Cobra antenna, 154-156
Coil, loading, 19, 113, 115-117
Cold flow, 89
Cold solder joint, 212
Conductor, metal, 21
Corrosion, 138-142

Counterpoise ground, 143, 198
Cross dipole antenna, 56, 64, 74
Cubical Quad antenna, 59-65, 164-172, 187-194
Current wave, 38
Cycle, sunspot, 29

"Dead spot", 149
Decibel, 24, 80
Demi-Quad antenna, 61, 164-168
Diamond Quad loop, 60
Dielectric, cable, 88
Dipole antenna, 21, 43, 70-73, 154
    radiation resistance, 37
    vertical, 44, 74
Directivity, 133
Director element, 24
Directional antenna, 22, 42, 51-55
    (see antenna, beam)
Dish antenna, 11
Drip loop, coaxial line, 94
Dummy load, 100, 199

Electric field, 14, 33
Electrical length, antenna, 19
Electrolytic corrosion, 138-140
Electromagnetic spectrum, 12, 13
Electromagnetic wave, 15, 33
Element, parasitic, 23, 53
End effect, antenna, 19
Extended vertical antenna, 47, 48

Feed point impedance, 34, 91
Ferrite r-f choke, 154, 170, 193, 200-203
Fiberglas, use of, 65, 113, 132, 191
Field, 5-7, 14, 33
"Figure-8" antenna pattern, 45
Figure of merit, Q, 113, 114
Filter, TVI, 206-208
Five-eighths wave antenna, 47, 48, 74
Foam dielectric cable, 91
Frequency, resonant, 17, 119
Front-to-back ratio, beam, 24

Gain, antenna, 24-26, 66, 75-80, 82, 83
Galvanic corrosion, 140-142
Gamma match, 203-205
Grid-dip oscillator, 119

Ground, counterpoise, 143
Ground, coaxial line, 197
Ground plane antenna, 45, 70-74, 145, 150-154
Ground wave, 28
Gutter clamp mount, 128
Guy wires, 149, 151

Half-power points, 69
Half-wave antenna, 49
    (see dipole)
Harmonic signal, 205
Helix antenna, 132
Hertz, 6, 12
Horizontal antenna, 44
Horizontal polarization, 32

Impedance, antenna, 35-40, 91, 104, 105
Incident wave, 102
Inductive reactance, 115
Input impedance, 34
Installation, mobile antennas, 111-134
Installation, coaxial plugs, 94-97
Interference, TVI, 205-208
Insulator, 87
Ionosphere, 7, 29, 30
Isotropic antenna, 24, 67-80

Lightning protection, 146, 196
Line of sight, 28
Lines of force, 5
Line impedance, 91
Line, coaxial, 87-101, 122
    (see coaxial line)
Load, antenna, 22, 47
Load, dummy, 100, 199
Loading coil, 19, 47, 113-115, 125
Lobe, antenna, 72
"Long John" Yagi antenna, 184-186
Loss, coaxial cable, 89, 101, 209

Magnetic field, 14
Marine antennas, 135-148
Marine ground, 144
Marine, lightning protection, 146
Matching, antenna, 22, 35-40, 91, 104, 105

Matching transformer, 176, 193, 200
(see balun)
Maxwell's Laws, 6, 35
Meter, SWR, 102-110
Mini-whip antenna, 113-129
Mobile antennas, 111-134
Monster Quad beam, 62, 80-82, 187-194, 219
Mismatch, coaxial cable, 35-40, 91, 104, 105
Multiple whip antennas, 131

National Electric Code, 198
Node, antenna pattern, 16
Noncontaminating vinyl, 90

Ohm's Law, 35
Omnidirectional antenna, 24, 44, 69
Oscillator, grid-dip, 119

Parasitic element, 23
Pattern, antenna, 50-54, 133
Phase, antenna, 40, 131
Plug, coaxial, 94-97
Polarization, 32, 51, 52, 62, 167, 172
Polar diagram, 69, 72
Power gain, 24, 66, 69, 78-80
Propagation, radio, 28

Q-Factor, 113, 114
Quad antenna, 59-65, 80-82, 187-194
Quad, Demi-, 61, 164-168
Quad, multi-element, 61-64, 168
Quad, Monster, 187-194, 219

Radiation resistance, 34-37, 91, 112, 114
Radials, 45, 75, 143, 154
Radio ground, 198
Radio range, 26, 133, 136
Radio wave, 33
Ratio, front-to-back, 24
Ratio, power, 24, 80
Reactance, 115
Reciprocity theory, 14
Reflected wave, 102
Reflectometer, 102-110
Reflector element, 24
Reradiation, antenna, 21
Resistance, coaxial line, 92
Resonance, antenna, 15-17, 21, 119
RTV-102 sealant, 93, 139

S-meter, 80, 85
Sailboat antenna, 136, 144
Silicone sealant, RTV-102, 93, 139
Skip distance, 28
Soldering techniques, 210, 211
Spectrum, electromagnetic, 12, 13
Sporadic-E layer, 30
Stacked antennas, 58, 59
Standing wave ratio (SWR), 16, 103
Sunspot cycle, 29
Surge impedance, 34, 91
SWR meter, 102-110

Television interference, 205-208
Transformer, balun, 63, 176, 193, 200-203
Transmission line, 87-101
(see coaxial line)
Trunk mount, antenna, 126
Truth Table, 78-80
Tune-up, SWR meter, 106-109
Turnstile antenna, 156-160

UHF coaxial fitting, 94
Unidirectional antenna, 23, 42, 51
(see antenna, beam)

Vertical polarization, 32, 44, 47
Vinyl, noncontaminating, 90

Waterproofing coaxial line, 93
Wave, current, 38
electromagnetic, 33
incident, 102
polarization, 32
reflected, 102
standing, 16, 103
Wavelength, 12, 16, 18
Whip antenna, 47, 111, 118-120, 131

Yagi antenna
all-metal array, 56
beam, 50-59, 173-186
gamma match, 203-205
"Long John", 184-186
multi-element, 173-184, 186
stacked, 58, 59